CROP PRODUCTION SCIENCE IN HORTICULTURE SERIES

This series examines economically important horticultural crops selected from the major production systems in temperate, subtropical and tropical climatic areas. Systems represented range from open field and plantation sites to protected plastic and glass houses, growing rooms and laboratories. Emphasis is placed on the scientific principles underlying crop production practices rather than on providing empirical recipes for uncritical acceptance. Scientific understanding provides the key to both reasoned choice of practice and the solution of future problems.

Students and staff at universities and colleges throughout the world involved in courses in horticulture, as well as in agriculture, plant science, food science and applied biology at degree, diploma or certificate level will welcome this series as a succinct and readable source of information. The books will also be invaluable to progressive growers, advisers and end-product users requiring an authoritative, but brief, scientific introduction to particular crops or systems. Keen gardeners wishing to understand the scientific basis of recommended practices will also find the series very useful.

The authors are all internationally renowned experts with extensive experience of their subjects. Each volume follows a common format covering all aspects of production, from background physiology and breeding, to propagation and planting, through husbandry and crop protection, to harvesting, handling and storage. Selective references are included to direct the reader to further information on specific topics.

Titles Available:
1. **Ornamental Bulbs, Corms and Tubers** A.R. Rees
2. **Citrus** F.S. Davies and L.G. Albrigo
3. **Onions and Other Vegetable Alliums** J.L. Brewster
4. **Ornamental Bedding Plants** A.M. Armitage
5. **Bananas and Plantains** J.C. Robinson
6. **Cucurbits** R.W. Robinson and D.S. Decker-Walters
7. **Tropical Fruits** H.Y. Nakasone and R.E. Paull
8. **Coffee, Cocoa and Tea** K.C. Willson
9. **Lettuce, Endive and Chicory** E.J. Ryder
10. **Carrots and Related Vegetable Umbelliferae** V.E. Rubatzky, C.F. Quiros and P.W. Simon
11. **Strawberries** J.F. Hancock
12. **Peppers: Vegetable and Spice Capsicums** P.W. Bosland and E.J. Votava
13. **Tomatoes** E. Heuvelink
14. **Vegetable Brassicas and Related Crucifers** G. Dixon
15. **Onions and Other Vegetable Alliums, 2nd Edition** J.L. Brewster
16. **Grapes** G.L. Creasy and L.L. Creasy
17. **Tropical Root and Tuber Crops: Cassava, Sweet Potato, Yams and Aroids** V. Lebot
18. **Olives** I. Therios
19. **Bananas and Plantains, 2nd Edition** J.C. Robinson and V. Galán Saúco
20. **Tropical Fruits, 2nd Edition Volume 1** R.E.Paull and O. Duarte
21. **Blueberries** J. Retamales and J.F. Hancock

CUCURBITS

2nd Edition

Todd C. Wehner

North Carolina State University
Department of Horticultural Science
Raleigh, North Carolina 27695-7609
USA

Rachel P. Naegele

USDA ARS, SJVASC 9611 Parlier,
California 93648
USA

James R. Myers

Oregon State University
Department of Horticulture
Corvallis, Oregon 97331-7304
USA

Narinder P.S. Dhillon

World Vegetable Center East and
Southeast Asia Kasetsart University
Kamphaeng Saen, Nakhon Pathom 73140
Thailand

Kevin Crosby

Texas A&M University
Department of Horticultural Sciences
College Station, Texas 77843-2133
USA

CABI is a trading name of CAB International

CABI
Nosworthy Way
Wallingford
Oxfordshire OX10 8DE
UK

Tel: +44 (0)1491 832111
Fax: +44 (0)1491 833508
E-mail: info@cabi.org
Website: www.cabi.org

CABI
WeWork
One Lincoln Street
24th Floor
Boston, MA 02111
USA

Tel: +1 (617)682-9015
E-mail: cabi-nao@cabi.org

A catalogue record for this book is available from the British Library, London, UK.

ISBN-13: 978 1 78639 291 6 (paperback)
       978 1 78639 292 3 (ePDF)
       978 1 78639 293 0 (ePub)

Commissioning Editor: Rachael Russell
Editorial Assistant: Emma McCann
Production Editor: Tim Kapp

Typeset by SPi, Pondicherry, India
Printed and bound in the UK by Bell & Bain Ltd, Glasgow

CONTENTS

PREFACE

Scientists attempt to make the world more understandable, enjoyable and usable for all. Horticulture is an important practical science with a long history. Although humans have been doing informal crop improvement for 12,000 years, scientific plant breeding did not begin until the late 17th century. Centuries of evolution have produced a diverse plant kingdom, requiring more detailed studies of distinct parts of the plant world to maximize our knowledge and use of individual crops.

The Cucurbitaceae is one of the most genetically diverse groups of crops in the plant kingdom. As a family and as individual crops, cucurbits epitomize adaptive differentiation and evolutionary divergence. Not only may cultivars within a crop vary significantly in their characteristics, but also the same cultivar grown in distinct areas can have different needs in response to diverse local growing conditions. Different cultures and ethnic groups have different cultivar preferences and horticultural practices, and that in turn increases the morphological diversity within the crop.

Crop production science has moved from general to specific. Uniformity in practice is being replaced by individual treatment of each small area of a farmer's field, study of the microenvironment, and the concept of stability of performance for cultivars. In research, we must now account for a greater number of the variables present in any biological experiment.

Research on cucurbits around the world has greatly increased our knowledge of this crop group. We could not begin to cover all the pertinent information in one text. Instead, we hope that this book will give the reader a general awareness of cucurbit crop production, an understanding of the underlying biological concepts as they pertain to cucurbits, and a jumping-off point from which to pursue investigations on particular crop species.

Although diverse, most cucurbits do share a collection of characteristics (e.g. rapidly growing vines with tendrils, possessing relatively large fruit, adaptable to the point of becoming weeds, containing various bioactive compounds) that make them a unique, fascinating, and useful family of plants. Continued research should lead to their enhanced exploitation and appreciation.

We would like to thank the previous authors of this book, R.W. Robinson and D.S. Decker-Walters, who formed the foundation on which we built this version. We would also like to thank Liberty Hyde Bailey, Tom Whitaker, Henry Munger, Clint Peterson, Carroll Barnes, Greg Tolla, Fred Andrus, Jim Crall, Gary Elmstrom, Don Maynard, Tom Williams, Linda Wessel-Beaver, Jim McCreight, Harry Paris, Brent Loy, V.S. Sheshadri, Tom Zitter and Michel Pitrat, who did so much to increase our knowledge and appreciation of cucurbits.

WHAT ARE CUCURBITS?

DISTRIBUTION AND ECOLOGY

The Cucurbitaceae is a family of frost-sensitive and predominantly tendril-bearing vining plants that are found in subtropical and tropical regions around the globe. There are only a few species that are native to temperate climates; they are either prolific seed-producing annuals, perennials that live for one season until killed by frost, or xerophytic perennials whose succulent underground parts survive the winter. Ecologically, the family is dichotomous; many genera flourish in the humid tropics, particularly in southeastern Asia and the neotropics, whereas other genera are native to the arid regions of Africa, Madagascar and North America. Members of the latter group, the xerophytes, usually have large perennial roots and succulent stems that are clambering and creeping and at least partially subterranean; in some cases, tendrils or leaves are lacking or greatly modified.

Although most crops in the Cucurbitaceae have been selected from the mesophytic annuals, concern over famine and fuel sources in arid countries has led to interest in turning some of the xerophytes into agricultural crops.

NOMENCLATURE

'Cucurbits' is a term coined by Liberty Hyde Bailey for cultivated species of the family Cucurbitaceae. Beginning in the early 20th century, the term has been used not only for cultivated forms, but also for any species of the Cucurbitaceae, and it will be so used in this book.

Other vernaculars applied to the family and various of its members are 'gourd', 'melon', 'cucumber', 'squash' and 'pumpkin'. Of these, squash and pumpkin are the most straightforward, almost always referring to species of *Cucurbita*. An exception is the fluted pumpkin, which is *Telfairia occidentalis*. The unqualified terms melon and cucumber usually define *Cucumis melo* and

Cucumis sativus, respectively. However, confusion develops when modifiers are added to these terms. Whereas muskmelon or sprite melon refers to a specific type of *C. melo*, watermelon is *Citrullus lanatus*, wintermelon is *Benincasa hispida* and bitter melon is *Momordica charantia*. Gourd generally is used to describe a cucurbit fruit with a hard, durable rind; usually it refers to bottle gourd (*Lagenaria siceraria*), or a wild species of *Cucurbita*, or an ornamental form of *Cucurbita pepo*. However, various other cucurbits also are called gourds, such as luffa sponge gourd (*Luffa aegyptiaca*), ridge gourd (*Luffa acutangula*) and some that do not have hard rinds, such as bitter gourd or bitter melon (*M. charantia*) and ivy gourd (*Coccinia grandis*). Sometimes the term refers to tough-rinded species of other plant families, such as the tree gourd (*Crescentia cujete* L., Bignoniaceae).

Complicating matters further, more than one common name is often applied to a single species. For example, names for *B. hispida* include ash gourd, white pumpkin, wax gourd and winter melon. Sometimes different names refer to distinct crops within a species, such as pumpkin, zucchini and acorn squash in *Cucurbita pepo*. In other cases, the different common names for the same species are used interchangeably. Frequently used common names for cucurbit crops are given in the Appendix.

Popular terms for the cucurbits may be confused in other languages as well. Cucumber and melon are sometimes considered the same in India. Older terminology in Chinese included 'guo-kua', which is translated into English as 'melon', but it includes watermelon (*C. lanatus*) as well as melon (*C. melo*). Also, 'tsaikua', which usually translates as 'gourd', refers to *B. hispida*, *C. sativus*, *M. charantia* and species of *Cucurbita* and *Luffa*. This is no longer a problem in China, since the names in Mandarin are now precise.

TAXONOMY

The family Cucurbitaceae, which is not closely related to any other plant family, consists of two well-defined subfamilies, eight tribes (representing varying degrees of circumscriptive cohesiveness) and about 118 genera and 825 species (Jeffrey, 1990). The four major cucurbit crops (watermelon, cucumber, melon, squash) and five other important crops (luffa, bottle gourd, chayote, wax gourd, bitter gourd) in the family belong to the Cucurbitoideae subfamily. Four of these – watermelon, luffa, bottle gourd and wax gourd – belong to the tribe Benincaseae. The classification of these and other cultivated species is given in Table 1.1. Many more wild taxa have actual or potential economic value, making Cucurbitaceae one of the most important plant families for human exploitation.

Taxonomic studies of cucurbits at all hierarchical levels have been done. They include comparative analyses of morphology (including specialized studies on trichomes, stomata, palynology and seed coat anatomy), cytology,

Table 1.1. Taxonomy of cultivated cucurbit species.

Latin name	Common name[a]	Frequency and place of cultivation[b]	Usage
Tribe Actinostemmateae			
Actinostemma tenerum Griff.	He-zi-cao, Goki zuru	Localized (S)	Medicinal
Bolbostemma (syn. *Actinostemma*) *paniculatum* (Maxim.) Franq.	Pseudo-fritillary	Localized (S)	Medicinal
Tribe Benincaseae			
Acanthosicyos horridus Welw. ex Hook. f.	Butterpips, Naras	Localized (A)	Food, medicinal, ornamental
Benincasa fistulosa (Stocks) Schaef. & Renner (formerly *Praecitrullus fistulosus*)	Round melon	Localized (W)	Food
Benincasa hispida (Thunb.) Cogn.	Wax gourd, winter melon	Frequent (W)	Food, medicinal
Citrullus colocynthis (L.) Schrad.	Colocynth	Sporadic (W)	Medicinal
Citrullus lanatus (Thunb.) Matsum. & Nakai.	Watermelon	Common (W)	Food, medicinal
Coccinia abyssinica (Lam.) Cogn.	–	Localized (A)	Food
Coccinia grandis (L.) Voigt	Ivy gourd	Sporadic (W)	Food, medicinal
Cucumis anguria L.	West Indian or burr gherkin	Localized (W)	Food
Cucumis dipsaceus Ehrenb. ex Spach	Teasel gourd	Sporadic (W)	Ornamental
Cucumis melo L.	Melon	Common (W)	Food
Cucumis metuliferus E. Mey. ex Naud.	African horned melon	Sporadic (W)	Food
Cucumis sativus L.	Cucumber	Common (W)	Food
Diplocyclos palmatus (L.) C. Jeffrey	Lollipop climber	Localized (O)	Ornamental
Lagenaria siceraria (Molina) Standl.	Bottle gourd	Common (W)	Utilitarian, ornamental, food
Melothria sphaerocarpa (Cogn.) Schaef. & Renner (formerly *Cucumeropsis mannii*)	White-seeded melon, Egusi-itto	Localized (A)	Food

Continued

Table 1.1. Continued.

Latin name	Common name[a]	Frequency and place of cultivation[b]	Usage
Tribe Bryonieae			
Bryonia alba L.	White bryony	Sporadic (W)	Medicinal, ornamental
Bryonia cretica L.	Bryony	Localized (O)	Medicinal
Bryonia dioica Jacq.	Red bryony	Sporadic (W)	Medicinal, ornamental
Ecballium elaterium (L.) A. Rich.	Squirting cucumber	Sporadic (O)	Ornamental
Tribe Cucurbiteae			
Cayaponia kathematophora R. E. Schult.	–	Rare (N)	Ornamental
Cayaponia ophthalmica R. E. Schult.	–	Rare (N)	Medicinal
Cucurbita argyrosperma C. Huber	Squash, pumpkin	Localized (W)	Food
Cucurbita ficifolia Bouché	Malabar gourd	Localized (W)	Food
Cucurbita ecuadorensis H. C. Cutler & Whitaker	Squash, pumpkin	Localized (N)	Food
Cucurbita maxima Duchesne	Great pumpkin	Common (W)	Food
Cucurbita moschata Duchesne	Squash, pumpkin	Common (W)	Food
Cucurbita pepo L.	Squash, pumpkin	Common (W)	Food
Sicana odorifera (Vell.) Naudin	Casabanana	Sporadic (N)	Food, ornamental
Tribe Gomphogyneae			
Gynostemma pentaphyllum (Thunb.) Mak.	Jiao-gu-lan, Doloe	Localized (S)	Medicinal
Hemsleya amabilis Diels	Luo-guo-di	Localized (S)	Medicinal
Tribe Joliffieae			
Telfairia occidentalis Hook. f.	Fluted pumpkin	Localized (A)	Food
Telfairia pedata (Sims) Hook.	Oyster nut	Localized (A)	Food

Tribe Momordiceae			
Momordica angustisepala Harms	Sponge plant	Rare (A)	Utilitarian
Momordica balsamina L.	Balsam apple	Frequent (W)	Medicinal, ornamental
Momordica charantia L.	Bitter gourd	Common (W)	Food, medicinal
Momordica cochinchinensis (Lour.) Spreng.	Cochinchin gourd	Sporadic (W)	Medicinal, ornamental
Momordica cymbalaria Hook. f.	–	Rare (O)	Food
Momordica dioica Roxb. ex Willd.	Kaksa	Localized (S)	Food
Tribe Sicyoeae			
Cyclanthera brachybotrys (Poepp. & Endl.) Cogn. (*Cyclanthera*, incl. *Rytidostylis* & *Pseudocyclanthera*, incl. *C. explodens* synonym of *C. brachystachya*)	Springgurka	Localized (W)	Food
Cyclanthera pedata (L.) Schrad.	Achocha, Stuffing cucumber	Sporadic (W)	Food, medicinal
Echinocystis lobata (Michx.) Torr. & Gray	Wild cucumber	Sporadic (W)	Ornamental
Hodgsonia macrocarpa (Bl.) Cogn.	Lard plant	Infrequent (S)	Food, medicinal, ornamental
Luffa acutangula (L.) Roxb.	Angled loofah	Frequent (W)	Food, medicinal
Luffa aegyptiaca Mill. (formerly *L. cylindrica*).	Smooth loofah	Common (W)	Utilitarian, food, medicinal
Sechium edule (Jacq.) Swartz	Chayote	Common (W)	Food, medicinal
Sechium tacaco (Pittier) C. Jeffrey	Tacaco	Localized (N)	Food
Trichosanthes costata Blume (formerly *Gymnopetalum cochinchinense*)	–	Rare (S)	Food
Trichosanthes cucumerina L.	Snake gourd	Frequent (W)	Food, medicinal
Trichosanthes dioica Roxb.	Pointed gourd	Localized (S)	Food, medicinal

Continued

Table 1.1. Continued.

Latin name	Common name[a]	Frequency and place of cultivation[b]	Usage
Trichosanthes kirilowii Maxim.	Chinese snake gourd	Localized (S)	Medicinal
Trichosanthes lepiniana (Naud.) Cogn.	Indreni	Localized (S)	Medicinal
Trichosanthes pilosa Lour. (formerly *T. ovigera*)	Japanese snake gourd	Frequent (S)	Food
Trichosanthes villosa Blume	Mi-mao-gua-lou	Localized (S)	Food
Tribe Siraitieae			
Siraitia grosvenorii (Swingle) A. M. Lu & Zhi Y. Zhang	Buddhafruit, Monk fruit, Luo-han-guo	Localized (S)	Food, Medicinal
Tribe Thladiantheae			
Thladiantha dubia Bunge	Red hail stone, Chi bao	Sporadic (W)	Medicinal, ornamental
Tribe Triceratieae			
Fevillea cordifolia L.	Antidote vine	Rare(N)	Medicinal

[a]For most species, the common name given is English or an English translation. Common names in other languages are listed in Kays and Silva Dias (1995) or US NPGS GRIN Taxonomy (2017).
[b]Locations: A=Africa, N=neotropics, O=Old World, S=Asia, W=widespread.

DNA, isozymes, flavonoids, cucurbitacins, amino acids and fatty acids in seeds, biogeography and coevolving insects. A monograph has been written on *Cucumis* (Kirkbride, 1993), and the taxonomic relationships within the Cucurbitaceae are being improved using molecular markers (Schaefer *et al.*, 2009; Chomicki and Renner, 2014). These new studies have been useful to the areas of crop improvement and germplasm conservation.

MORPHOLOGY AND ANATOMY

Seedlings

Most cucurbit seedlings are epigean, germinating with the tips of the cotyledons initially inverted but later erect. As it emerges, the hypocotyl straightens and the cotyledons ascend as the seed coat is dislodged by the peg, an outgrowth on one side of the hypocotyl. The function of the peg is to open the seed coat and permit the cotyledons to emerge. The photosynthetic cotyledons of most cucurbit seedlings are more or less oblong in shape. Between them lies the inconspicuous developing epicotyl.

Roots

Cucurbits generally have a strong taproot, which may penetrate the soil to a depth of more than 2 m, as in the case of squash. Even in cucumber, the tap root can extend 1 m into the soil. Cucurbits also have many secondary roots occurring near the soil surface. In fact, most roots are in the upper 60 cm of the soil. Lateral roots extend out as far as, or farther than, the above-ground stems. The cortex of the primary root is apparently involved in the development of secondary roots in cucurbits. Adventitious roots may arise from stem nodes in squash, luffa, bitter gourd and other cucurbits, sometimes without the stem having contact with the soil or other substrate.

Some xerophytic species have massive storage roots that enable the plant to survive severe drought. Those of *Acanthosicyos* can reach up to 15 m in length. The central taproot on one buffalo gourd plant weighed 72 kg (Dittmer and Talley, 1964). The above-ground parts of this species may die from lack of water or in response to freezing temperatures, but the plant regenerates from surviving stem tissue at the root–shoot transition area when favourable conditions resume.

Older vessels in the secondary xylem are often plugged with tyloses (extensions of the parenchyma cells), especially in watermelon, which can contribute to drought resistance. The sieve tubes in the secondary phloem are among the largest found in angiosperm plants.

Stems

The herbaceous or sometimes slightly woody stems are typically prostrate, trailing, or climbing, angled in cross-section, centrally hollow, sap-filled and branched. Primary and secondary branches can reach 15 m in length. Bush forms of cucurbits have much shorter internodes as well as total stem lengths than vining cultivars.

Many of the xerophytic cucurbits are true caudiciforms; that is, the lower part of the perennial stem, which is usually subterranean or at ground level, is thickened, succulent and drought resistant. In *Marah*, the large underground tubers originate from the hypocotyl and stem base (Stocking, 1955). The succulent stems of *Ibervillea sonora* (S. Wats.) Greene can continue to sprout new growth annually during periods of drought lasting 8 years or more (Macdougal and Spalding, 1910).

The vascular bundles of cucurbit stems are bicollateral (phloem to the inside and outside of the xylem), discrete, usually ten, and arranged in two rings around the pith cavity. The relatively large sieve tubes are also scattered in the cortex in some cucurbits (e.g. squash), serving to join all phloem elements together. The anomalous stem anatomy of cucurbits and other vines may serve to increase stem flexibility, to facilitate nutrient transport, to promote healing of injuries, or to provide protection against stem destruction via redundancy (Fisher and Ewers, 1991).

Many cucurbits have soft to rough hairs (trichomes) on their stems and foliage, whereas chayote, smooth luffa, stuffing cucumber and some other cucurbits are glabrous or nearly so. Trichome morphology is quite variable: hairs are glandular or eglandular, unicellular to multicellular, and simple or branched.

Leaves

Cucurbit leaves are usually simple (i.e. not divided into leaflets), palmately veined and shallowly to deeply three- to seven-lobed. There is usually one leaf per stem node. Along the stem, leaves are helically arranged with a phyllotaxy of 2/5; in other words, there are two twists of the stem, which segment contains five leaves, before one leaf is directly above another. This means that the angle of divergence between neighbouring leaves is 144° (2/5 of 360°).

Leaf stomata are mostly anomocytic, lacking subsidiary cells. The petiole in cross-section often has a crescent or ring of unequal vascular bundles, the larger ones bicollateral. Stipules at the base of the petiole are typically absent, but have been transformed into photosynthetic thorns in *Acanthosicyos*. Extrafloral nectaries, which frequently attract ants, occur on some cucurbit leaves (e.g. ivy gourd).

Succulent leaves are rare, even among the xerophytic cucurbits. Those of *Xerosicyos* have large water-storage cells in the inner mesophyll and perform crassulacean acid metabolism (CAM). However, the deciduous leaves of *Seyrigia* only perform C_3 photosynthesis even though the succulent stem performs CAM. The large ephemeral leaves of most cucurbits adapted to an arid environment avoid heat damage by maintaining high levels of transpirational cooling (Rundel and Franklin, 1991).

Tendrils

Most cucurbits have solitary tendrils at their leaf axils. Tendrils are unbranched in species such as cucumber and branched in luffa and other taxa. They are often coiled, helping plants to cling to trellises and other supports. Terminal adhesive pads develop on the tendrils of several species, allowing attachment to tree trunks and other large textured objects. Some cucurbits lack tendrils, e.g. squirting cucumber and bush cultivars of summer squash, while other cucurbits may have more than one tendril per node.

Tendrils in most of the cucurbit crops are interpreted as modified shoots. However, in luffa and other species, they are considered a stipule–stem complex. There are still other interpretations concerning the anatomical origins of cucurbit tendrils and research is ongoing. In cucumber, a single nucleotide polymorphism (SNP) resulted in the transition from tendrils to tendril-less, removing the plant's ability to climb (Wang *et al.*, 2015).

Flowers

Many cucurbits have large showy flowers that attract pollinating insects, but *Echinocystis*, *Sechium* and some other genera have small, rather inconspicuous flowers. The typically unisexual flowers occur in leaf axils, either alone or in inflorescences. They are often white or yellow, but may be red (e.g. *Gurania*) or other colours. The hypanthium is cup- or bell-shaped. The sepals or sepal lobes, typically numbering five, and the corolla, which is usually five-lobed and more or less fused, extend beyond the hypanthium. Flowers have radially symmetrical, bell-shaped corollas that may differ between male (staminate) and female (pistillate) flowers.

Staminate and pistillate flowers on monoecious cucumber and squash plants are originally bisexual, with both stamen and pistil primordia initiated. During ontogeny, depending on the hormonal status of the tissue near the floral bud, development of the anthers may be arrested and a pistillate flower develops, or development of the pistil is retarded and a staminate flower is produced. Undeveloped stamens (staminodia) can be seen in mature pistillate flowers, and there is a rudimentary pistil (pistillodium) in staminate flowers (Fig. 1.1).

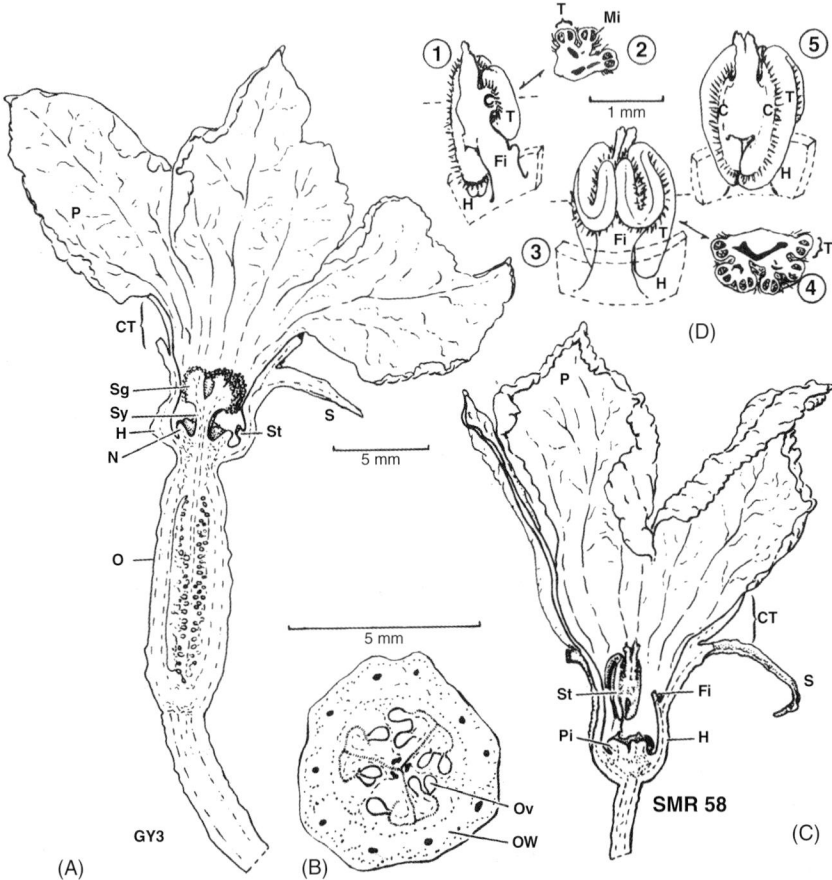

Fig. 1.1. Comparison of female and male cucumber flowers. (A) Female flower at anthesis, longitudinal section. (B) Tricarpellate ovary, transection at anthesis. (C) Male flower at anthesis, longitudinal section. (D) Simple stamen (1, 2) and two compound stamens (3–5). C, anther connective; CT, corolla tube; Fi, filament; H, hypanthium; Mi, microsporangium; N, nectary; O, ovary; Ov, ovule; OW, ovary wall; P, petal; Pi, pistillodium; S, sepal; Sg, stigma; St, stamen or stanimodium; Sy, style; T, theca. (Goffinet, 1990. Reprinted courtesy of Cornell University Press.)

Stamens are attached to the hypanthium and alternate with corolla lobes. The basic number of stamens in the Cucurbitaceae is five. Some cucurbits (e.g. *Fevillea*) have five free stamens, whereas all five stamens are fused together in *Cyclanthera*. During evolution, fusion of two pairs of stamens has resulted in some genera (e.g. *Cucumis*) having one small unilocular and two large bilocular stamens. The three stamens are usually attached to each other to some degree by their anthers, as in squash; filaments are short and often united also.

In the Cucurbitaceae, the subfamily Zanonioideae has relatively homogeneous pollen morphology. In the subfamily Cucurbitoideae, pollen is quite variable, but consistent within genera. For example: in squash, pollen grains are very large, spherical and spiny, but in cucumber the pollen grains are more globular and smooth.

Pistillate flowers have an inferior ovary below the hypanthium. The pistil, which often has a fused style but separate stigma lobes, generally has three or five carpels. The fleshy placentae bear numerous ovules in most species, but only one in chayote.

Floral nectaries attract pollinating insects. These structures are borne inside and at the base of the hypanthium in both staminate and pistillate flowers. The nectary forms a continuous ring surrounding the base of the style(s) in the female flower, whereas the nectary and its associated pistil rudiment form a button-shaped mound at the centre of the male flower.

Fruit

Fruit of the Cucurbitaceae are extremely diverse in many characteristics, including size, shape, colour and ornamentation. Those of bryony are small (ca 5 mm), spherical, and green, red or black in rind colour. Angled luffas are club-shaped, about 60 cm long, and prominently ribbed. Some of the many shapes of bottle gourds are described in Chapter 4. The striped, mottled, bicoloured or solid-coloured fruit of squash are smooth, wrinkled, warted, furrowed or ridged. *Sicyos* has stiff, dry spines, whereas those on the teasel gourd are soft and fleshy.

Cucurbita maxima is well named, for it is this species that is the giant of the plant kingdom. Every year, there is a contest to grow the world's largest pumpkin. In this contest any squash fruit with an orange skin is considered a pumpkin, and the winners are invariably *C. maxima*. Weights crossed the 500 kg threshold in 1999 and the 1000 kg threshold in 2014. The winning fruit in 2016 weighed an astounding 1193 kg. Giant watermelon contests are also run, with the winning fruit in 2015 weighing 137 kg.

Cucurbit fruit are generally indehiscent 'pepos', usually with one or three ovary sections or locules (Fig. 1.2). A pepo is a fleshy fruit with a leathery, non-septate rind derived from an inferior ovary. However, fruit of some cucurbit genera, e.g. *Momordica* and *Cyclanthera*, split at maturity. Fruit of squirting cucumber forcefully eject their seeds through a blossom-end pore. Fruit may be dry when mature, as in luffa, where seeds fall out through a hole at the bottom of the pendulous fruit. Mature fruit of many cucurbits have a hard, lignified rind, but various squash cultivars have been bred to have a tender rind.

Cucurbit fruit flesh is generally white to pale yellow and moist, but cultivated cucurbits such as melon, squash and watermelon have been bred to have a range of flesh colours. Melon and squash have also been bred to include green

Fig. 1.2.   Frontal and cross section of female flower from *Cucurbita pepo* (spaghetti squash).

or orange flesh, and watermelon flesh can be white, salmon yellow, canary yellow, orange, coral red or scarlet red. In melon, watermelon and squash, the flesh is derived from the fruit wall. In other species, including cucumber, the edible flesh may be mostly placental in origin.

Many cultivated cucurbits produce a fleshy fruit at maturity. Others, such as gourd and luffa, dry at maturation. When a luffa fruit matures and dries, what is left is a papery outer skin and a fibrous mass surrounding the seeds. This tangled mass of modified vascular bundles, consisting mostly of fibre cells, makes up a luffa sponge.

Seeds

There may be only one seed in the fruit, as in chayote, or more typically, tens to hundreds of seeds. Cucurbit seeds, which are rarely winged, are usually flat. The seed coat encloses a collapsed perisperm, an oily embryo and little or no endosperm. The tiny endosperm is consumed during seed development. Two cotyledons make up much of the contents of the seed. Seeds of some cucurbits are enveloped in a false aril of placental origin; in bitter gourd this sarcotesta is red and fleshy, attracting birds as seed dispersal agents.

Seed size, shape and colour vary greatly among the cultivated cucurbits (Fig. 1.3). The largest unwinged seed is that of *Hodgsonia*, measuring about 7 cm long. The nearly spherical seed of *Bryonia* is sometimes less than 3 mm in diameter. Cultivars within a crop, including watermelon and squash, may differ considerably in their seed sizes and other seed characteristics. Depending on the cultivar, watermelon seeds are white, tan, brown, black, red, or green. They can also have patterns on them, referred to as dotted, rimmed (dark seed margin), tipped (dark seed tip) or clump (dark seed centre).

The complex seed coat anatomy of cucurbits has been well studied (Singh and Dathan, 1990). The testa develops from the outer integument of the

Fig. 1.3. Seeds of the Cucurbitaceae. The largest seed (*Fevillea cordifolia*) shown here measures ca 53 mm across. Also shown are seeds of *Cucurbita, Siraitia, Marah, Momordica, Sicana, Trichosanthes, Luffa, Lagenaria, Echinocystis* and *Melothria* (the smallest seeds).

anatropous, bitegmic, crassinucellate ovule. The inner integument degenerates in fertilized ovules. The mature seed coat in the subfamily Cucurbitoideae consists of an epidermis, hypodermis, main sclerenchymatous zone, aerenchymatous zone and inner parenchymatous or chlorenchymatous zone. In the Zanonioideae, the sclerenchymatous layer is poorly or not differentiated from the hypodermis, and the well-developed aerenchyma has distinctive lignified thickenings. Within each subfamily, there is further anatomical diversity among genera.

GROWTH AND DEVELOPMENT

Seed germination

Seed maturation usually continues until the fruit starts to yellow with senescence. Some seed producers store mature cucurbit fruit after harvest to permit the seeds to develop further. However, if seeds are left too long in some fruit, they may germinate *in situ*. In a few cucurbits, such as chayote, germination is naturally viviparous.

Seed dormancy, which is common in various wild species, is not usually a serious problem in the major crops. Dormancy can occur in freshly harvested seeds of some cultivars, but this dormancy can be broken by a month or more of after-ripening, i.e. storing seeds in the fruit after harvest. Light and low temperature ($< 15°C$) are strong inhibitors of germination for many species. Under amenable conditions (e.g. low light levels, temperatures of $25–30°C$ and adequate but not soaking moisture), germination takes 2 days to 2 weeks if the seeds are not dormant. See Chapter 6 for experimental studies investigating seed germination physiology.

Plant growth and movement

Most cucurbits grow rapidly in warm weather, with stem extension growth outpacing leaf development in the tuberous perennials. In a single growing season, a wild buffalo gourd plant produced 360 shoots covering an area 12 m in diameter with a total vine length of over 2000 m (Dittmer and Talley, 1964). Among the annuals, bottle gourd stems can elongate up to 60 cm in 24 h, and wild cucumber (*Echinocystis lobata*), which is adapted to the short growing seasons of southern Canada, is considered one of the fastest growing vines. Holroyd (1914) reported that a single annual squash (*C. pepo*) plant produced 450 leaves on a vine measuring 43 m. Cucurbit root growth is also rapid, occurring at a rate of 6 cm per day for squash when conditions are favourable. Elite cultivars of pickling cucumber will go from seed planting to first harvest in 39 days if grown at the optimum growth temperature (32°C).

In large-fruited vining squash plants, total leaf area increases exponentially throughout the season until fruit set creates a large reproductive sink and vegetative growth is suppressed. During the period between flower primordia initiation and the start of fruit development in West Indian gherkin, differentiation of new vegetative organs decreases, the growth rate of existing vegetative organs increases, and water and nutrient intake drops; soon after fruit development begins, vegetative differentiation and water and nutrient intake resurge (Hall, 1949).

Biomass productivity differs among species and cultivars and is influenced by cultural practices (e.g. planting density, irrigation, fertilizer application) as well as local environmental conditions. Environmental factors most affecting growth rates are photoperiod and ambient temperature, either of which can affect the intake and effective utilization of water and nutrients. Many investigations of the relationship between growth and environmental factors have been carried out on species of *Cucumis*, particularly cucumber. The results of these studies, which were discussed in detail in Whitaker and Davis (1962) and Wien (1997), are summarized below.

1. The growth rate curve for a single leaf under continuous light is generally an S curve, but is affected by light intensity.
2. The rate of stem elongation is greater during 8 h days than during 16 h days, and plants grown under short-day conditions produce more nodes and leaves, but smaller total leaf and root areas.
3. Overall stem length may be greater under a long-day versus a short-day regime when nitrogen levels are high.
4. Under low-nitrogen conditions, plants grown during long days contain more carbohydrates at anthesis than plants grown during short days, but this carbohydrate relationship is reversed at fruit maturity.
5. Stem extension and leaf area growth rates are linearly dependent on mean ambient temperature during periods of optimum temperatures for growth

(20–30°C, depending on other environmental conditions). However, in the range of 15–27°C, Grimstad and Frimanslund (1993) found that plant dry weight of cucumber had a sigmoidal response curve with inflections at about 17°C and 24°C.

6. When temperature rises above the optimum, leaf growth rate in young plants declines as material is redistributed to the stems, and cell division in developing leaves is reduced.

7. At below-optimum temperatures, relative leaf growth rate is independent of temperature and is controlled instead by light intensity.

8. Stem extension rates are lower than normal when night temperatures exceed day temperatures.

9. Low temperatures slow the development of apical buds.

In *C. pepo*, bushy plants with short internodes possess an allele that reduces biosynthesis of endogenous gibberellin. When these plants are treated with a high concentration (2.9–4.3 mmol l^{-1}) of gibberellic acid, internode lengths become as long as those of naturally viny squash plants.

Breeders have also selected for bush cultivars in *C. maxima*. Research on bushy versus viny plants of this species indicates that bush cultivars have a more uniform growth pattern, respond better to high-density planting, and produce a greater percentage of fruit versus vegetative biomass during short growing seasons (Loy and Broderick, 1990). This last effect is partly due to the fact that fruiting begins sooner in bush cultivars, which in turn suppresses vegetative growth. However, photosynthetic rates in bush plants increase during fruit set in response to increasing sink demand, which may be possible because of a proportionally thicker palisade layer in bush plant leaves (Loy and Broderick, 1990).

In climbing cucurbits, the stems often revolve, twist and extend upwards. Darwin (1906) made several interesting observations on the revolving nature of cucurbit stems and tendrils. He noted that the average rotation rate was 100 min per revolution in wild cucumber. Light affects this movement, with stem tips, including the two uppermost internodes subtending the apical meristem, following the sun throughout the day.

Generally, the slightly curved tendril becomes sensitive to mechanical stimulus on its concave side when it is one-third grown. At that stage, it reacts quickly (in under 2 min), with coiling caused by the elongation of parenchymatous cells on the convex side of the tendril. The revolving movement of a tendril does not stop after it has coiled, but its ability to coil is limited after it stops revolving.

Sex expression

No cucurbit species is known to have only or primarily functionally hermaphrodite flowers. Instead, most cucurbits are monoecious; that is, they have separate male and female flowers on the same plant. Among its genera, the Cucurbitaceae also has a high rate of dioecy, where staminate and pistillate flowers occur on separate plants. Some genera have only dioecious species. Half of the 135 cucurbit species surveyed in India by Roy and Saran (1990) were dioecious, a much higher proportion than is typical for angiosperm families. Cultivated cucurbits that are dioecious include oyster nut, fluted pumpkin, ivy gourd, pointed gourd and monk fruit (luo-han-guo, *Siraitia grosvenorii*).

Primitive cucurbits are typically monoecious, as are the majority of domesticated cucurbits. Most squash and watermelon cultivars are monoecious, although genes for different forms of sex expression are known for these crops. For example, some watermelon cultivars may have three types of flowers on the same plant: staminate, pistillate and perfect (hermaphrodite). The wild-type watermelon is andromonoecious (staminate followed by perfect flowers). Many cucumber cultivars are monoecious, but others are gynoecious or predominantly gynoecious, and round-fruited cucumber such as 'Lemon' and 'Crystal Apple' are andromonoecious. All seven sex types have been identified in cucumber: androecious, gynoecious, hermaphroditic, monoecious, andromonoecious, gynomonoecious, and trimonoecious (staminate, perfect and pistillate flowers). Angled luffa is monoecious with some exceptions, such as the cultivar 'Satputia', which has only hermaphrodite flowers. Most round-fruited melon cultivars are andromonoecious, with the fruit formed from perfect flowers. The bisexual flowers of these plants are often borne on the first or second node of the lateral branches. There are exceptions, however. Monoecious cultivars such as 'Athena' have been selected to have short, almost round fruit developing from pistillate flowers. 'Banana' and other melon cultivars of the Flexuosus Group are monoecious. The group includes snake melon (Armenian cucumber) and pickling melon, with long fruit having crisp white bland flesh, similar to cucumber.

Monoecy is the ancestral condition in cucurbits. Dioecy and other forms of sex expression have arisen in various evolutionary lines in the family. Single genes can determine the occurrence of unisexual plants in normally monoecious species, as in the case of dioecy (all male or all female flowers on a plant) in *C. pepo*. In melon, cucumber and many other cucurbits, two or more genes are involved in sex expression, sometimes (as in *Luffa*) with each gene having three or more alleles. The development of heteromorphic sex chromosomes (e.g. ivy gourd) to determine dioecy is considered to represent the ultimate degree of evolution from monoecy in the Cucurbitaceae.

Monoecious cultivars may differ in degree of female sex expression, some having a higher proportion of female to male flowers. Generally, they produce many more staminate than pistillate flowers and go through a progression of floral development. Nitsch *et al.* (1952) determined that young squash plants are initially vegetative, then bear only underdeveloped male flowers. Later, they produce only normal male flowers, then bear normal female as well as male flowers. The proportion of female to male flowers increases as the plant grows older, and the plant eventually produces only female flowers. If squash plants are not pollinated, ultimately they may produce enlarged female flowers and parthenocarpic fruit.

Although staminate flowers usually appear a few days to a few weeks before pistillate flowers on squash vines, a few cultivars and some wild populations of *C. pepo* may produce female flowers first (Fig. 1.4), especially when spring temperatures are low. These wild populations also tend to have a higher ratio of female to male flowers than most cultivars (Decker, 1986).

In their search for early staminate flower production in cucumber, Walters and Wehner (1994) examined 866 cultivars, breeding lines and plant introduction accessions. Earliness was normally distributed for the cultigens and ranged from 26 to over 45 days from planting to first staminate flower.

Sex expression in the Cucurbitaceae is controlled by environmental as well as genetic factors. Unfavourable growing conditions (e.g. lack of water) can reduce flower production. In general, female sex expression is promoted by low temperature, low nitrogen supply, short photoperiod and high moisture availability, i.e. conditions that encourage the build-up of carbohydrates. These environmental factors influence the levels of endogenous hormones (especially ethylene, auxin and gibberellic acid), which in turn influence sex expression. Most studies examining this role of hormones have been conducted on

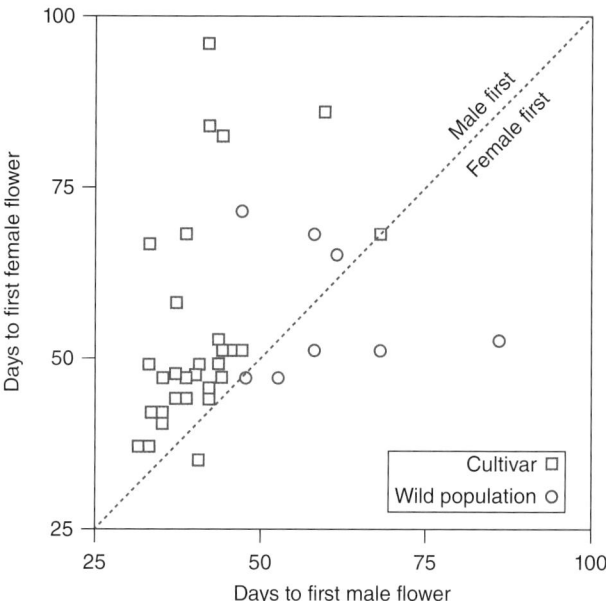

Fig. 1.4. Accession means of the number of days from seed germination to the first male and female flowers on plants of 30 cultivars and eight wild populations of *Cucurbita pepo*. (Data from Decker, 1986.)

cucumber, followed by squash, melon and watermelon. See Chapter 7 for details on the use of exogenous hormones to control sex expression.

Temperature affects anthesis and, in squash, the length of time that a flower is open. Squash pollen is released at temperatures as low as 10°C, whereas cucumber, watermelon and melon flowers require higher temperatures for anther dehiscence. Warmer temperatures cause anthesis in squash to occur earlier in the morning. However, high temperatures (ca 30°C) accelerate squash flower closing, causing the corollas to close by mid- to late morning.

Fruit development

Pollen tube growth and ovule fertilization stimulate ovary enlargement. Subsequent fruit set depends on the quality of pollination (i.e. having enough ovules fertilized) and is affected by the presence of already developing fruit, leaf area, daylength and other environmental factors. Fruit on a plant may inhibit the production of additional pistillate flowers and the development of subsequent fruit. In cucurbit crops such as melon, watermelon and squash, fruit-thinning will allow fruit that remain on the vine to grow larger.

If fruit are not developing on a plant by the end of the growing season, then the last group of ovaries may develop parthenocarpically. Parthenocarpy in cucumber and squash is promoted by low temperature, short daylength, old plant age and genetics. Some cucumber cultivars have genes for parthenocarpy and will set fruit without pollination. Parthenocarpic cultivars are common in the greenhouse trellis type, the Middle Eastern greenhouse (Beit Alpha) type and, more recently, field slicers and field pickles. Parthenocarpic pickling type has become popular in northern Europe and most of the USA, due to seedless fruit and higher yield. Field production of parthenocarpic cultivars depends on isolation from conventional cultivars having staminate flowers, as well as the exclusion of beehives from the area.

Anderson (1894) determined that a developing squash fruit gained weight at an average rate of 1 g per minute. The greatest weight increase was at night. The growth rate for cucurbit fruit is influenced by exogenous conditions (e.g. higher temperatures and greater light exposure increase the growth rate) as well as by endogenous plant conditions, such as the presence of other developing fruit, which retards growth.

Several studies on the inheritance and development of fruit shape in squash (*C. pepo*) were conducted by Sinnott (1932). He reported that fruit shape is evident in the shape of the immature ovary, with ultimate shape affected by both genetic and environmental factors. Also, fruit that are set first may be shaped differently than those set later, the difference being evident in the shape of the ovaries.

Immature ovaries are usually green, although those of squash cultivars with gene *B* may be yellow. Fruit of various cucurbits, including luffa and bottle gourd, remain green until fruit senescence, at which time they turn tan or brown. Other cucurbits develop rind coloration changes during maturation. In these, chlorophyll depletion reveals the presence of additional pigments after pollination. For example, the green fruit of many squash cultivars become yellow or orange as they age. Colour changes usually begin at the blossom end of the fruit. In most wild and cultivated *Cucurbita*, rind patterns, such as stripes or mottling, are lightly visible on the ovary, becoming more distinct soon after pollination. However, the white fruit of 'Mandan' (*C. pepo*) reach almost full size before the dark green markings appear. Fruit markings persist at maturity for some squash cultivars, but fade away in senescent fruit of others.

In young melon fruit of the Cantalupensis Group of cultivars, rapidly dividing cork cells develop below the epicarp. Near maturity, this growth breaks through to form a network of grey corky tissue covering the rind, as is evident in the netting of muskmelon cultivars.

A unique case of adaptation to a particular ecological niche is exemplified by *Cucumis humifructus* Stent. This African species has a geocarpic fruit, similar to that of peanut. After flowering and setting fruit above ground, similar to other *Cucumis* species, the developing fruit is thrust downward and completes its development several inches below the soil surface. In its native land, the African anteater or aardvark (*Orycteropus afer*) consumes the subterranean fruit. It is a symbiotic relationship, with the aardvark using the fruit as a source of food and water in the arid area where the plant often grows, and providing a means of seed dispersal for the plant. *C. humifructus* is called aardvark cucumber because of this relationship, but it is more closely related to melon than to cucumber.

CROP EVOLUTION AND DIVERSITY

EVOLUTIONARY HISTORY

'Cucurbits are weeds waiting to become crops.' While this anthropomorphic statement may be an over-simplification of crop evolution in the Cucurbitaceae, supporting evidence can be seen in the vast array of valuable plant products found within the family. Aboriginal plant gatherers were probably attracted to some of these products, particularly the relatively large, long-keeping and sometimes showy fruit. After fruit were taken back to camp, seeds that were purposely discarded, accidentally dropped or partially digested found new life on rubbish heaps, settlement edges or other disturbed areas near camp. Eventual recognition of the value of the resident cucurbits led to their appreciation, horticultural care and further exploitation. Finally, seeds and, more rarely, vegetative propagules were carried by and exchanged among migrating bands of these incipient cultivators, gradually turning the earliest cultivated cucurbits into domesticated crops.

It was not only because cucurbits were wanted by human gatherers that they became domesticated. Certain physiological and genetic characteristics generally associated with weeds, or colonizing species, allowed cucurbits to adapt to human habitats. Fast, indeterminate growth, developmental plasticity in response to environmental conditions (especially regarding sex expression) and genetic diversity at the genomic, chromosomal and gene levels enabled these cucurbits to continue their survival through the coevolutionary relationship that humans call domestication. In turn, this relationship brought many changes to domesticated cucurbits: the size of various plant parts increased; early female flower production was selected; seed dormancy was reduced; photoperiod response was eliminated; fruit bitterness was eliminated; and fruit flavour and appearance were improved.

Given the nature of cucurbits, it is not surprising that they are among the most ancient of cultivated plants. Archaeological evidence suggests a pantropical distribution for bottle gourd going back more than 10,000 years.

Bottle gourd may have been cultivated in Asia, Africa and the New World at that time, or later if the oldest remains are of wild plants. Using molecular taxonomy, even older times can be seen, although humans did not exist then. Evidence of human use of cucurbits in the form of seeds or fruit rinds of *Cucurbita pepo* has been found at sites dating ca 10,000 BP in Mexico and ca 5000 BP in central USA (Smith, 1997a, 2006). Phytoliths are particles of silicon dioxide found in the cells of *Cucurbita* spp. These particles remain in soil samples and can be used as a diagnostic tool in archaeological studies of this genus. This technique has been used to document the presence of domesticated *Cucurbita* spp. in South America as early as 12,000 BP (Piperno and Stothert, 2003). Enlarged fruit parts indicate that other squash species were being cultivated for food in the New World by 7000 BP, if not earlier.

Cucumber (*Cucumis sativus*) is of Asiatic origin, with its primary centre of diversity in India and secondary in China. Although cucumber evolved from melon, the two species cannot be crossed. Cucumber can be crossed with *Cucumis sativus* var. *hardwickii*, *C. s.* var. *sikkimensis* and *C. s. xishuangbannanensis*. Also, *Cucumis hystrix*, a wild species with traits from both melon and cucumber, has been crossed successfully with cucumber. However, *Cucumis setosus* is a melon relative that looks like cucumber, but will not cross with it. Cucumber has been grown in Asia for several thousand years. The remains of *Cucumis* crops (melon or cucumber) in eastern Iran have been dated to the third millennium BC.

Melon (*Cucumis melo*) likely originated in India from *C. trigonus* or *C. callosus* (Sebastian *et al.*, 2010). The Australian species *C. picrocarpus* has the highest DNA sequence similarity to *C. melo*. However, the oldest wild relatives of melon, such as *C. hirsutus*, *C. humifructus* and *C. sagittatus*, originated in Africa. Melon was grown in Egypt in the second millennium BC and in Iran during the third millennium BC. Humans moved the melon throughout the Middle East and Asia, making it an important vegetable in India, Egypt, Iran and China. Iran, Afghanistan and China are secondary centres of melon diversification. Spain is a tertiary centre of diversity.

Melon was one of the most important vegetables in ancient China, with archaeological remains there dating back to 5000 BP. The origin and domestication of melon is still debated, though evidence for two lineages have been described, with an Indian relative (*C. trigonus*) being the most closely related living relative to *C. melo* (Endl *et al.*, 2018).

Watermelon (*Citrullus lanatus*) is an African native, and became important in northern Africa and southwestern Asia by 6000 BP. It is recorded in the Bible (Numbers 11:5) that the Jews missed the watermelons and cucumbers of Egypt during the Exodus in the 12th century BC. Probably, that passage refers not to cucumber but to snake melon, also known as Armenian cucumber. A watermelon relative, colocynth (*Citrullus colocynthis*), is probably the toxic fruit mentioned in the Bible in II Kings.

USES

The uses of cucurbits are diverse. Fruit may be eaten when immature (e.g. summer squash) or mature (watermelon). They are baked (squash), fried (bitter gourd), boiled (wax gourd, snake gourd), stuffed (stuffing cucumber), pickled (cucumber, pickling melon), candied (watermelon, squash, wax gourd, Malabar gourd), consumed fresh in salads (cucumber) or as a dessert (melon, watermelon). Fresh fruit are sold soon after harvest or following a storage period of several weeks to months. Processing for long-term storage often involves canning, freezing, or pickling (Fig. 2.1). Juice from cucumber fruit is used to make salad dressing, lotion and shampoo. Watermelon fruit juice is used directly as a fresh drink, but results in a 'squash flavour' when pasteurized. Juice from the fruit of watermelon, Malabar gourd and other cucurbits is fermented to make alcoholic beverages; for example, a liqueur is made from melon in Japan. Fruit are the most important plant part that is used. However, seeds (*Cucurbita*, *Cucumeropsis* and many others), roots (chayote), flowers (squash) and leaves and shoot tips (*Telfairia*, *Momordica*) also provide human sustenance.

Seeds of watermelon, oyster nut, !nara and other cucurbits are an important part of the African diet. Watermelon seeds are used to make snacks, using the large (long) seed size (not the medium, short or tomato seed sizes), usually with a white seed coat. In Mexico and Central American countries, squash seeds are sold in stores and at street corners as snacks. The development of squash and pumpkin cultivars with 'naked' seeds, devoid of tough seed coats, has increased the popularity of squash seeds for food in other countries.

Fig. 2.1. Pickled snake melon.

Cucurbits are used for other purposes besides food and drink. From bottle gourds, people have fashioned storage containers, bottles, cups, bowls, utensils, smoking pipes, musical instruments, penis sheaths, masks, floats for fishnets, rattles for babies and many other items. The dry fruit rinds of other cucurbits, including wax gourd, wild and cultivated species of *Cucurbita* and fluted pumpkin, sometimes serve as containers and utensils as well. The oils extracted from seeds of watermelon, squash, luffa, antidote vine, lard plant, white-seeded melon, colocynth and other cucurbits are prepared for cooking, illumination, candle and soap manufacture and industrial purposes. Colocynth seeds have 18% oil (70% of which is linoleic). The fibrovascular system of a mature luffa fruit provides a sponge suitable for various purposes, including use as a filter. Nigerians pound and work the thick stems of sponge plant to yield white, absorbent fibres that serve as a washing sponge. Stem fibres of fluted pumpkin are extracted for use as a sponge and for making rope. Jewellery is fashioned from seeds of various cucurbits (*Cayaponia kathematophora*, *Marah* species). Squash flowers are used in make-up and cucumber fruit extracts are added to soap and shampoo. Arabs employ the dried fruit pulp of colocynth in the making of gunpowder, tinder and fuses.

Medicinal applications of cucurbits are numerous. Since ancient times, indigenous cultures worldwide have employed cucurbits to treat ailments. A Chinese medicinal text written during the 1st century AD mentions the therapeutic benefits of wax gourd, pickling melon, bottle gourd and Japanese snake gourd. In modern Chinese commerce, he zi cao (*Actinostemma lobatum*), jiao-gu-lan (*Gynostemma pentophyllum*), luo-han-guo (*Siraitia grosvenorii*) and species of *Trichosanthes* are the most important medicinal cucurbits (Yang and Walters, 1992; Cai *et al.*, 2004; Xu, 2017).

Some of the reported therapeutic properties, including the use of cucurbits as purgatives, emetics and anthelmintics (anti-parasite drugs), are due to cucurbitacins, which are bitter triterpenoid compounds known to have drastic effects on the digestive system. Cucurbitacins are present in almost all cucurbits, usually throughout the plant, but especially in fruit and roots. Other toxic and potentially medicinal compounds in cucurbits include saponins, free amino acids and alkaloids.

Cucurbit fruit, and their extracts, are the plant part most commonly prepared for medicine. The gourd-like fruit of colocynth is used as a purgative for constipation and parasitic worms, as an anti-tumour agent for cancer and as a remedy for fever, urogenital disorders and other problems. The bitter roots of many cucurbits (e.g. bryony) have cathartic effects as well. Leaves are employed for medicine less frequently, often dried and pounded into a powder or made into a tea (*G. pentophyllum*). Seeds of squash, wax gourd, bottle gourd and other cucurbits are taken as anthelmintic medicine. Oil extracted from seeds of antidote vine (*Fevillea cordifolia*) is used medicinally to counteract the poison of a snakebite, to ward off dandruff and for an assortment of other ailments.

One of the most widely employed of the medicinal cucurbits is bitter gourd. Research has shown that this plant serves to control, though not cure, mild to moderately chronic cases of diabetes mellitus by increasing carbohydrate utilization. Medical investigators in China, Japan and India are testing species of *Momordica* for their purported analgesic (pain relief), abortifacient (causing abortion), immunosuppressive and anti-tumour properties. Similar laboratory research is ongoing for species of *Trichosanthes* (Rai *et al.*, 2009; Liyanage *et al.*, 2016). In 1989, compound Q (related to methane monooxygenase), which is an active principle extracted from Chinese snake gourd, received a lot of attention when medical professionals began to study it as a treatment against the human immunodeficiency virus (HIV). Bitter gourd has also been used to combat HIV infection.

The bioactive compounds in cucurbits also serve other purposes. People in Mexico have long taken advantage of the saponins in fruit of wild *Cucurbita* for creating a frothy cleansing soap. Similarly, Nigerians use the leaves of balsam apple to clean metals, and the leaves and fruit as a body wash. Saponins in seeds and fruit make an effective component in fish poison. For example, luffa is used for that purpose in Australia. Fruit of casabanana are hung in the house as a room deodorizer. Various cucurbits are used to repel insects. The Chinese spray extracts of *Luffa* and *Momordica* on crops to control spider mites and other agricultural pests (Yang and Tang, 1988). In North Africa, Arabian camel herders smear a rind extract of colocynth on water bags to keep camels away.

Cucurbit fruit sometimes become objects of decoration and art. In the USA, ornamental gourds and turban squashes provide autumn table decorations, and pumpkins (*C. pepo*) are carved and illuminated with candles to celebrate the holiday of Halloween, as well as celebrations of autumn harvest. The ultimate degree of artistic expression has been reached with bottle gourds. Many different cultures independently developed the custom of carving, painting and otherwise decorating the gourds that were so useful to them. The art of gourd decoration is most highly developed in Peru, where detailed pyro-engraving techniques produce fine-art gourdcraft exported around the world.

Finally, some cucurbits are grown as ornamental plants. In addition to its lacy leaves and colourful fruit, the fragrant blossoms of bitter gourd make this species a very appealing trellis ornamental. Bitter gourd has been grown in glasshouses in the UK since Victorian times. The beautiful white, fringed petals of snake gourd and its long serpent-like fruit, dangling beneath a garden arbour, make for a picturesque sight. The spiny fruit of teasel gourd and African horned melon and the explosive fruit of squirting and exploding cucumbers have brought these cucurbits into commercial trade as well. The most recent horticultural trend affecting cucurbits is the passion for succulents. Collectors are beginning to learn of and trade among themselves the caudiciform cucurbits of Africa (e.g. *Gerrardanthus* and *Trochomeria*), Madagascar (*Seyrigia* and *Xerosicyos*) and North America (*Marah* and *Ibervillea*).

COMMERCIAL PRODUCTION AREAS AND IMPORTANCE

According to the Food and Agriculture Organization (FAO) of the United Nations, watermelon is the most popular cucurbit in the world (Table 2.1). The FAO estimated that 3,477,000 ha of watermelon were harvest worldwide in 2014, producing 111,009,000 t of fruit. Next in total world production were cucumber, melon, and then squash and pumpkin. All of these crops have seen increases in production since the 1970s (Table 2.1). The greatest growth in harvested area (355%) was seen in pumpkins and squashes, but an approximate 200% increase was also seen for other cucurbit crops. In watermelon, production has increased by 444% over the past three decades. Cucumber production has seen a 343% increase in production over the same time period.

China remains the leading producer of the major cucurbit crops in the world. Although China is also a prominent producer of luffa, wax gourd and other cucurbits of Asian origin, production statistics for these crops are not readily available. China exports various cucurbit products, including fresh fruit, watermelon and squash seeds, and dried fruit of the medicinal (and low-calorie sweetener) monk fruit or luo-han-guo (Fig. 2.2). In 1990, the Chinese Agricultural Department estimated that 30% of the land in China that was devoted to vegetable production was sown in cucurbits (Yang and Walters, 1992). In 2014, the second leading producer of watermelons and melons was Turkey. Russia was the second leading producer of cucumbers, and India was second for squash and pumpkins (Table 2.1).

It is very difficult to obtain reliable statistics on cucurbit production in many countries because governmental statistics are not readily available to the public and data gathering and reporting procedures vary from country to country. Also, different agencies report different statistics for the same country.

Besides the major crops, there are some minor cucurbit crops of international commercial importance. Already mentioned are the fruit of monk fruit (luo-han-guo) exported from China. China and Japan both export other cucurbits of medicinal interest, such as preparations of *Gynostemma* and *Trichosanthes*, which are finding their way into the American herbal medicine market. Medicinal preparations of colocynth have a long history of trade from African to European pharmacies. Fruit of African horned melon that are commercially produced in New Zealand appear in American and European food markets sporting the trade name 'kiwano'. Because of expanding demand, Kenya and Israel have also become important exporters of this attractive fruit. Japan and Central America are the major exporters of luffa sponges. In the early 1940s, when luffa sponges were an important source of industrial filters, Brazil exported almost 2 million sponges per year to the USA alone. In 1985, Costa Rica was the leading exporter of chayote. In 2011, Mexico became the leading exporter of chayote to the USA with 19,500 t being exported, though

Table 2.1. Leading countries in the production of cucurbits.

Crop	Country	Area harvested (1000 ha)		Yield (kg ha⁻¹)[a]		Production (1000 tonnes)	
		1979–81	2014	1979–81	2014	1979–81	2014
Squash and pumpkin	China	70	392	10,513	18,468	737	7,241
	India	245	519	8,117	9,606	244	4,987
	Russia	–	60	–	20,490	–	1,232
	Ukraine	–	54	–	20,417	–	1,104
	USA	–	36	–	24,126	–	863
	World Total	564	2,004	10,328	12,573	5,825	25,197
Cucumber	China	395	1,178	11,841	48,235	4,678	56,855
	Russia	193	70	7,945	26,022	1,533	1,820
	Iran	57	89	12,217	20,129	693	1,804
	Turkey	32	66	15,968	26,977	503	1,780
	Ukraine	–	52	–	18,165	–	940
	World Total	1,093	2,179	12,726	34,414	13,908	74,975
Melon	China	89	439	17,120	33,613	1,528	14,753
	Turkey	75	101	20,783	16,904	1,567	1,707
	Iran	42	77	11,182	19,225	475	1,477
	Egypt	17	38	21,101	27,983	359	1,050
	India	–	46	–	22,350	–	1,034
	World Total	594	1,179	14,554	25,133	8,648	29,626
Watermelon	China	230	1,852	18,387	40,405	4,228	74,843
	Turkey	151	157	20,952	24,668	3,157	3,885
	Iran	158	133	10,713	26,871	1,700	3,568
	Brazil	–	94	–	23,009	–	2,171
	Egypt	51	70	23,453	28,775	1,107	2,015
	World Total	1,786	3,477	13,990	31,923	24,973	111,009

[a]World Yield totals are for average yield.
Source: FAOSTAT (2017). Values for 1979–81 are averages over those years.

Fig. 2.2. Luohan beverage mix from China. Seeds and fruit of luo-han-guo are pictured on the package.

Fig. 2.3. Bitter gourd, immature fruit and canned. (Reprinted by permission from The New York Botanical Garden, © 1988, *Economic Botany* 42, 287, J. Ruh and J. Thieret, photographers, in Walters and Decker-Walters, 1988.)

Costa Rica's production was similar. Both Costa Rica and Mexico still ship large amounts of chayote to other Latin American countries and to Europe. Decorated bottle gourds are a source of foreign currency in Peru. Snake gourd and bitter gourd (Fig. 2.3) are bottled or canned for export in Asia and have been for decades.

Cucurbits are important in the international seed trade. In addition to the major food crops, seed companies offer germplasm of other edible and ornamental cucurbits such as luffa, wax gourd, bottle gourd, bitter gourd, stuffing cucumber and casabanana.

CUCURBITS AS WEEDS

Many domesticated species can become troublesome weeds, particularly in agriculture and elsewhere. Feral populations of cucurbit crops occur along roadways, railroad tracks, stream banks, settlements and waste areas in the Americas, Europe, Africa, Asia, Australia and even islands in the Pacific. Particularly widespread and persistent are weedy forms of bitter gourd, watermelon, citron and species of *Cucumis*. Also common are escapes of species of *Cucurbita* and *Luffa*, ivy gourd, colocynth, bottle gourd, squirting cucumber and red hailstone (*Thladiantha dubia*).

Cotton fields in southern Asia and southeastern USA are plagued with weedy plants of Queen Anne's pocket melon (*Cucumis melo* var. *dudaim*) and Texas gourd (*Cucurbita pepo* ssp. *texana*). The gourd problem has affected soybean and cotton production in Mississippi, Louisana, Arkansas and Texas.

West Indian gherkin (*Cucumis anguria*) is reportedly a pest in sugar cane and peanut fields in Australia. This species also grows wild in many areas of the New World, including Brazil and various Caribbean islands. It is not native to the New World; rather, it escaped from cultivation, apparently after being introduced with the African slave trade.

Commercial seed trade has been another disperser of weeds. During the 19th century, species of Old World bryony (*Bryonia* spp.) were sold in the USA as ornamentals and sometimes escaped. Similarly, wild cucumber (*Echinocystis*), a native of North America, has escaped from European gardens.

3

BREEDING AND GENETICS

THE MECHANICS OF PLANT BREEDING

Manual pollination

Large unisexual flowers and sticky pollen facilitate manual pollination in cucurbit breeding experiments. In squash, male and female flowers to be used the next morning are chosen the previous day. The closed corolla of an appropriate squash flower will have a light touch of yellow at the apex. The corolla tips of the staminate flower may be slightly separated also (Fig. 3.1). The apices of the chosen flowers are taped, tied or covered to prevent pollinator entry. The next morning, a staminate flower is removed and opened and the dehisced pollen is rubbed on to the stigma of a previously tied female flower. After pollination, the female flower is bagged, or the corolla re-secured to prevent pollinator contamination, and the developing fruit is then tagged.

Because squash pollen does not store well, the use of fresh pollen is customary in squash breeding. However, pollen can be used from pre-anthesis flowers that are kept for a few days at low temperature and high humidity (Robinson, 2010–2011). Also, pistillate flowers can be pollinated 1 day pre-anthesis. Cucumber can be pollinated 24 h after the flower opens, especially if pollinations are being made in the greenhouse or growth chamber and if the temperature is below 30°C.

Cross-pollination of andromonoecious melons and watermelons by hand is more difficult than for monoecious cultivars, since the anthers need to be manually excised from each perfect flower. The anthers of perfect flowers are usually removed on the afternoon before anthesis. Afterwards, the emasculated flower is enclosed to prevent insect pollination. As with squash, the male flowers of melon are either closed before anthesis or picked and kept indoors overnight at high humidity. Pollen transfer from these flowers takes place the following morning.

Fig. 3.1. Cross-section of female and male flowers from *Cucurbita pepo*: (left) Spaghetti winter squash and (right) Patty Pan summer squash one day before anthesis, at which time the corollas are tied shut to exclude insects from flowers that are to be hand-pollinated.

Tissue culture

Tissue culture is used in cucurbits for embryo rescue, propagation, embryogenesis and organogenesis. Success in producing interspecific plantlets by embryo culture has been achieved in *Cucurbita* (*C. pepo* × *C. moschata*, *C. pepo* × *C. ecuadorensis*) and *Cucumis* (*C. metuliferus* × *C. zeyheri* Sond.). Other crosses, such as African horned melon and melon, have been difficult. So far, there are no reports of the successful use of protoplast fusion in creating interspecific hybrids of cucurbits.

Cultivars within a cucurbit crop often differ in their ability to be regenerated through tissue culture. Embryos have been obtained from various plant tissues (e.g. leaf callus, hypocotyl explants, protoplast-derived callus), but their development into plants has been sporadic. Cotyledons and hypocotyls have proved to be the best explant tissues for organogenesis, but leaves, fruit and embryos have also been used.

In their production of transgenic melon plants, Gonsalves *et al.* (1994) cultured the developing tissue on Murashige and Skoog medium according to plant developmental stage (i.e. regeneration, shoot elongation and rooting stages). They also performed regeneration experiments in which they found that proximal explants produced a higher percentage of regenerated shoots than distal tissue samples, with most shoots arising from the proximal side of the square-cut piece of cotyledon.

Achieving breeding objectives

Progress in cucurbit breeding has been improved through the use of multivariate statistical analyses (e.g. to increase selection efficiency), research on gene mapping and linkage, the induction of mutations with chemicals and gamma radiation, marker assisted selection, the use of wider crosses within species and gene transfer among species, and the collection and screening of germplasm in genebanks for desirable characteristics.

Although genetic engineering and backcrossing usually focus on single alleles, more attention is being given to multiple gene selections for single or multiple characteristics. For example, in Saudi Arabia, the melon cultivars 'Najd I' and 'Najd II' were bred for tolerance of both high temperature and salinity. In some cases where multiple insect resistance has been found, it is possible to breed for resistance to several insects from a single cross. 'Butternut' squash (*Cucurbita moschata*), for example, is resistant to squash vine borer, cucumber beetles, pickleworm, melonworm and leafminer. Squash bugs prefer *Cucurbita pepo* and *C. maxima*, but if these species are not present, then they will feed on *C. moschata*. Sources of combined insect and disease resistance are also known for cucumber and melon. Often, quantitative resistance to a disease provides a lower level of resistance, but resistance to multiple isolates of the pathogen. Genetically modified cultivars of squash for virus resistance have been combined with conventionally bred resistance to powdery mildew as well as with fusarium wilt in melon (Quemada and Groff, 1995).

Selection for many morphological characteristics is standard practice in plant breeding. Progeny are visually inspected for the desired character and further breeding proceeds according to objectives. The study of Mendelian genetics in families can help reveal relationships (e.g. dominant, co-dominant, recessive, complementary) among alleles at multiple loci.

Although biochemical qualities can be evaluated with the help of laboratory equipment and analysis (e.g. using a refractometer to select for high sugar content in fruit), linkage or pleiotropy for morphological characters has also been used to breed for these traits. For example, selection for fruit flesh carotenes was initially based on fruit colour, i.e. orange-fleshed fruit were presumed to have higher concentrations of these organic compounds. However, subsequent research in squash (*C. pepo*) revealed that fruit colour is controlled by the complex interaction of alleles at several genes, and although total carotenoid content is affected by these genes, there are additional genes influencing the percentage of carotenes in the total carotenoid content (Paris, 1994). Also, the major allele (*B*) for high carotenoid content has various pleiotropic, and not always favourable, effects on fruit and foliage. Direct selection for high carotene content has proved to be more effective. The *B* gene is not a factor in *C. maxima* and *C. moschata*, and selection for high carotenoid content and sometimes specifically for carotenes has been more successful in these species.

Gene linkage, and also pleiotropy, has been a problem in breeding monoecious melons. The use of monoecious melons in F_1 hybrid seed production is desirable because emasculation of the female parent is easier with the pistillate flowers of monoecious inbred lines than with the perfect flowers of andromonoecious inbred lines. However, selection for monoecy has been limited by its genetic association with elongated fruit shape. Elongated shape is dominant and F_1 hybrids with a monoecious parent generally have undesirably long fruit. Other genes can influence the shape of melons with the monoecious gene, and H.M. Munger of Cornell University was successful in breeding monoecious selections with nearly-round fruit.

Many objectives in cucurbit breeding involve the improvement of traits having complex inheritance. Earliness, yield, adaptation to certain environmental conditions and fruit quality are quantitative traits. Consequently, large populations, efficient experiment designs and multivariate analyses are useful for evaluation of breeding material for future selections. Recurrent selection and pedigree selection have been used to improve cucurbits for quantitative traits. Backcrossing has been used especially for improvement of sex expression and disease resistance (Munger *et al.*, 1993).

Studies with cucumber, melon, watermelon and squash (Whitaker and Davis, 1962; Wehner, 1999) indicate that, in general, there is little or no inbreeding depression, but there is some heterosis for certain morphological traits. Increasingly, wider crosses are being employed to produce F_1 populations with more favourable characteristics. The mechanics of F_1 seed production are described in Chapter 7. In any case, hybrids are useful for the following reasons.

1. Hybrids permit the protection of the parental inbreds by trade-secret laws, although it is still possible for the female line to appear occasionally in the hybrid population. Companies may seek additional protection with intellectual property rights (utility patent, plant variety protection, breeder's rights).
2. Hybrids can be quickly produced that have interesting combinations of the parental traits, such as intermediate fruit length from a long-fruited crossed with a short-fruited inbred.
3. Hybrids combine the dominant alleles from the female parent with the dominant alleles from the male parent to produce a cultivar with all of the alleles expressed together.
4. Hybrids can combine cytoplasmic traits such as chilling tolerance from the female inbred with dominant alleles such as chilling tolerance from the male inbred into a more tolerant progeny.
5. Seedless hybrids can be produced from certain interspecific combinations or, in the case of watermelon, seedless triploid hybrids can be produced by crossing tetraploid and diploid parents.

Interspecific hybridization

Augustin Sagaret and Charles Naudin tried to cross melon with cucumber in the mid-19th century, but without success. All later investigators trying to cross these distantly related species have also failed, but interspecific hybrids have been produced for other species of *Cucumis* (Fig. 3.2), as well as for *Cucurbita* (Fig. 3.3), *Citrullus*, *Luffa*, *Momordica*, *Trichosanthes* and other cucurbit genera.

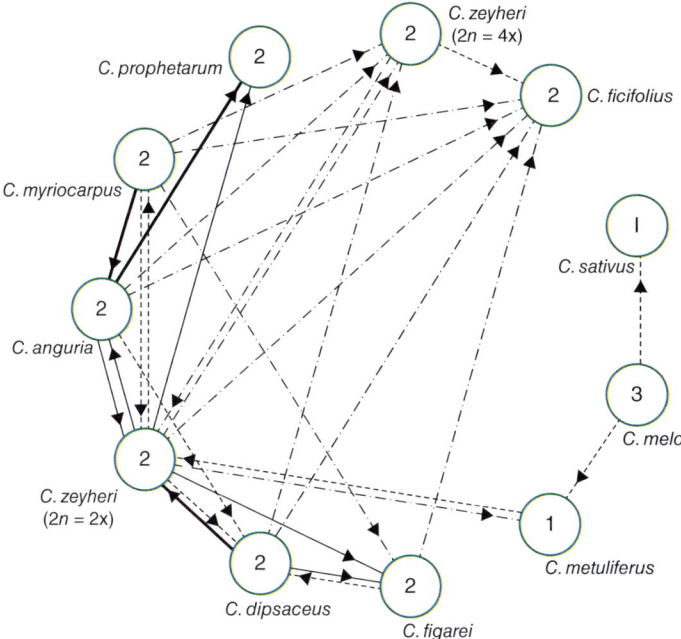

Fig. 3.2. Crossability polygon of *Cucumis* species. Arrows point in the direction of the female parent. Moderately to strongly self-fertile and cross-fertile hybrids (thick solid line); sparingly self-fertile and moderately cross-fertile hybrids (thin solid line); self-sterile, usually not cross-fertile hybrids (dash-and-dot line); inviable seeds or seedlings (dashed line). Absence of a line indicates that seeded fruit were not obtained. (Nijs and Visser, 1985. Reprinted courtesy of Kluwer Academic Publishers.)

Cucurbita is an example of where interspecific gene transfer has been utilized successfully for crop improvement. However, *Cucumis sativus* has been improved using genes from *C. sativus* var. *hardwickii*. Also, *C. sativus* was crossed with wild *Cucumis hystrix* and the chromosomes doubled to create a new allotetraploid, *C. hytivus*. *Citrullus lanatus* has been improved using *Citrullus mucosospermus*, *C. amarus* and *C. colocynthis*.

Although the production of interspecific hybrids is only the first step in a rather long process, F_1 hybrids of *Cucurbita maxima* × *C. moschata* are used directly to produce elite cultivars. Both parental species are monoecious, having many more male than female flowers, but the interspecific hybrid is gynoecious or predominantly gynoecious in sex expression. The interspecific hybrid is usually productive if conditions for pollination are favourable (e.g. a monoecious cultivar is grown nearby to provide pollen, and bees are working the field). The unusual case of a gynoecious hybrid being produced by crossing two monoecious species also occurs in the cross of *C. pepo* × *C. ecuadorensis*.

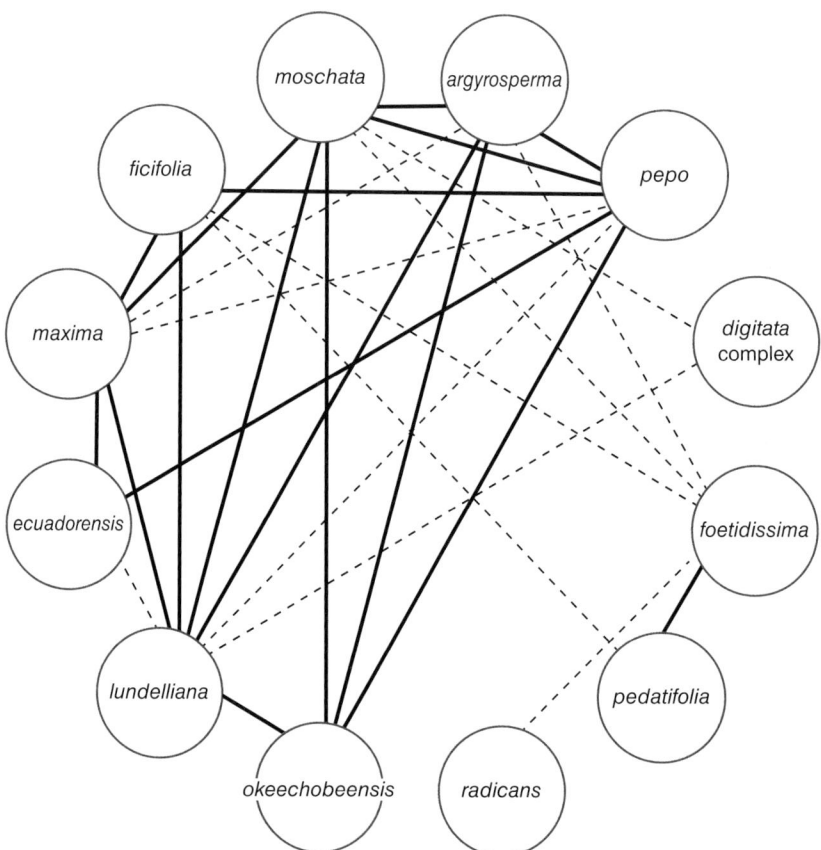

Fig. 3.3. Crossability polygon of *Cucurbita* species. 'Digitata Complex' includes *C. digitata, C. palmata, C. cylindrata* and *C. cordata*, which are considered subspecies of *C. digitata* by some scientists. All crossing combinations have been tried in at least one direction, except for *C. pedatifolia* with *C. maxima*, and *C. radicans* with *C. pedatifolia, C. ficifolia, C. ecuadorensis, C. okeechobeensis*, and *C. digitata sensu lato*. Early works describing hybridization attempts with *C. radicans* were in error as described in Merrick (1990); the material was misidentified. Other published sources, as well as the unpublished work of Tom Andres (New York, 1996, personal communication), were used to create this diagram. Solid lines indicate an F_1 hybrid that is at least partially fertile; dashed lines indicate a viable but sterile F_1 plant.

The cross *C. maxima* × *C. moschata* may be difficult or easy to make depending on parental combinations (Castetter, 1930; Yongan *et al.*, 2002; Karaağaç and Balkaya, 2013). For difficult crosses, many pollinations may be needed to set each fruit, and only a few seeds per fruit are produced. Breeders in Japan have been able to make this cross so successfully that they market interspecific F_1 seed commercially. They have found *C. maxima* and *C. moschata* parents that cross well and are more compatible than most members of these species. 'Tetsakabuto', the first popular interspecific hybrid squash cultivar, is a cross of 'Delicious' (*C. maxima*) × 'Kurokawa No. 2' (*C. moschata*). Seed production for the interspecific hybrid is most prolific when *C. maxima* is the maternal parent.

The cross *C. maxima* × *C. pepo* is difficult but not impossible to make; the F_1 is highly sterile. *C. pepo* and *C. moschata* can be crossed, but compatibility is influenced greatly by the choice of parents, with some cultivars crossing more readily than others. Wall and York (1960) reported that the cross was more easily made when one of the parents was an F_1 hybrid, which increased gametic diversity. The bush gene of *C. pepo* has been introgressed into *C. moschata* to incorporate compact plant habit into that species. Zucchini yellow mosaic virus (ZYMV) resistance, derived from 'Nigerian Local' (*C. moschata*), has been moved into *C. pepo* to produce resistant cultivars such as 'Tigress' and 'Jaguar'.

Of all the *Cucurbita* species, *C. moschata* and *C. argyrosperma* are the most closely related and hybridization may occur in nature, especially where these two species are sympatric in Central America (Wessel-Beaver, 2000). The interspecific cross is easy to make, although only with *C. moschata* as the male parent (Wessel-Beaver *et al.*, 2004). F_1 hybrids with commercial potential are under development in Mexico (Ortiz-Alamillo *et al.*, 2007).

Wild *Cucurbita* species are being used to develop disease-resistant squash. In disease studies, *C. ecuadorensis* and *C. foetidissima* (buffalo gourd) were found to be resistant to a greater number of viruses than other species of *Cucurbita* tested (Provvidenti, 1990). Buffalo gourd is difficult to use in squash breeding because of its distant relationship and incompatibility with the cultivated species (Fig. 3.3). However, virus resistance alleles have been successfully introgressed from *C. ecuadorensis* to *C. maxima*. Multiple virus-resistant germplasm derived from *C. maxima* × *C. ecuadorensis* was developed at Cornell University, USA, and provided to breeders in 1985. 'Redlands Trailblazer', a winter squash resistant to ZYMV, watermelon mosaic virus (WMV) and papaya ringspot virus (PRSV), was bred in Australia from the same interspecific cross. Cultivars of *C. pepo* vary considerably in their compatibility with *C. ecuadorensis* (Robinson and Shail, 1987). Distant interspecific crosses of *Cucurbita* can be made more successful if embryo culture is used.

Resistances to cucumber mosaic virus (CMV) and powdery mildew have been transferred from *Cucurbita okeechobeensis* to *C. pepo* and *C. moschata*. The cross *C. okeechobeensis* × *C. pepo* is difficult to make, even with embryo culture, and sterility is a serious problem with the interspecific hybrid. These difficulties are overcome by the use of a bridging species, for example crossing *C. pepo* to the F_1 of *C. okeechobeensis* × *C. moschata* (Whitaker and Robinson, 1986).

Because undesirable traits (e.g. bitterness) of a wild parent are often dominant, unacceptable horticultural types predominate in the F_2 generation of these interspecific crosses. However, only a single backcross to 'Butternut' (*C. moschata*) of the F_1 of 'Butternut' × *C. okeechobeensis* ssp. *martinezii* was needed to produce disease-resistant (powdery mildew and CMV) plants with fairly good fruit characteristics. In general, the number of backcrosses to the domesticated parent that are needed depends on the number of genes that are different between the two parents. Once acceptable progeny are produced, self-pollination is performed to obtain uniformity.

Wild species of *Cucurbita* could probably be used further in squash breeding programmes to provide resistance to other pathogens and insects. Other desirable traits, such as the drought tolerance and gynoecious sex expression of *C. foetidissima*, could be transferred as well.

CYTOLOGY

Chromosome numbers have been determined for most of the important cultivated species of the Cucurbitaceae and a number of other cucurbits since the 1990s (Table 3.1). The most common haploid numbers are 11 and 12, with 12 considered to be the ancestral karyotype number for the family. Squash species

Table 3.1. Chromosome number for some species of the Cucurbitaceae.

Species	Haploid number of chromosomes	Genome size (Mb)
Benincasa hispida	12	800
Citrullus lanatus	11	425
Coccinia grandis	12	719/904*
Cucumis anguria	12	–
Cucumis dipsaceus	12	–
Cucumis melo	12	454
Cucumis metuliferus	12	–
Cucumis sativus	7	367
Cucurbita species (e.g. *C. pepo*)	20	283
Cyclanthera explodens	16	–
Cyclanthera pedata	16	–
Lagenaria siceraria	11	313
Luffa species (e.g. *L. cylindrica*)	13	790
Momordica balsamina	11	–
Momordica charantia	11	339
Momordica cochinchinensis	14	–
Momordica dioica	14	–
Praecitrullus fistulosus	12	–
Sechium edule	12	–
Trichosanthes cucumerina	11	–
Trichosanthes dioica	11	–

*indicates that estimated genome size for female/male plants

have 20 pairs of chromosomes, more than any other cultivated species in the family, and cucumber has the fewest, with only seven pairs. The hypothesis that cucumber evolved by fusion of the chromosomes of melon was rejected in the 1980s, but later was proved to be true through genome sequencing (Huang *et al.*, 2009). Studies of synteny show that the 11 chromosomes of watermelon became the 12 chromosomes of melon, and finally became the seven chromosomes of cucumber (Huang *et al.*, 2009).

Chromosome morphology of cucurbits is difficult to study, because the chromosomes are small and not easily differentiated from the cytoplasm by cytological procedures. Fortunately, this can now be done using molecular marker technology. The cytology and phylogenetic relationships of many cucurbit crop species was studied in the 1980s and 1990s (Singh, 1990). Variation was observed for length of chromosomes, position of centromere, occurrence of satellites and karyotypes for cucumber, melon, *Momordica* species, *Luffa cylindrica* and *Lagenaria siceraria* (Ramachandran and Seshadri, 1986; Bharathi *et al.*, 2011; Waminal and Kim, 2012). Few karyotyping studies have been conducted in *Cucurbita*. Waminal *et al.* (2011) used fluorescence *in situ* hybridization (FISH) based on the 5S and 45S rDNA to karyotype *C. moschata*, but similar studies have not been conducted for other *Cucurbita* spp.

DNA content using flow cytometry (C-value) distinguishes cucurbit species and sexes within species (Achigan-Dako, 2008). In bottle gourd species, inter- and intra-variation has been used to differentiate seed type and genome size (Achigan-Dako *et al.*, 2008). Heteromorphic sex chromosomes have been identified in dioecious species of *Coccinia* and *Trichosanthes*, but are lacking in other dioecious cucurbits (Roy and Saran, 1990). In ivy gourd (*Coccinia grandis*), plants with two X chromosomes are female and XY plants are male. The Y chromosome is larger and more heterochromatic than the X chromosome and is dominant in sex determination. Polyploid plants with three X and only one Y chromosome are androecious. In some dioecious species (e.g. *Coccinia*), C-values have shown that the Y chromosome can increase genome content up to 10% (Sousa *et al.*, 2016).

Cucurbita is believed to be an ancient tetraploid genus derived from an ancestor with a base chromosome number of 12 (Wu *et al.*, 2017). Isozymic and sequencing evidence have implied a polyploid nature of this genus (Weeden, 1984). That polyploidy occurred long ago is indicated by all species of the genus having 20 pairs of chromosomes and by disomic 3:1 gene ratios in segregating generations. Recent genomic sequencing efforts provide evidence of genome duplication shortly after *Cucurbita* diverged from other cucurbit species ~30 million years ago (Montero-Pau *et al.*, 2017).

Aside from *Cucurbita* and possibly other genera of the tribe Cucurbiteae, polyploidy does not seem to have played an important role in the evolution of cucurbit tribes or genera. However, isolated cases of polyploid species are known in some genera with mostly diploid species (e.g. *Cucumis* and *Trichosanthes*) and polyploid cytotypes occasionally arise in melon and a few other species (e.g. kaksa).

Autotetraploids have been induced in cucumber, melon, squash, water-melon, luffa and bottle gourd by treatment with colchicine. Other compounds such as Surflan (oryzalin) herbicide can be used to produce tetraploids. Tetraploids have not been used directly in horticulture so far, except in the production of seedless triploid watermelons.

Within-species triploids have been created in cucumber, melon, squash and watermelon by crossing tetraploids with diploids. Triploids are highly sterile, whereas tetraploids are more fertile than triploids but less fertile than diploids. The sterility of triploids, due to embryo abortion, has been used to produce seedless watermelon (see Chapter 4 for more details). Although the use of triploidy in breeding other cucurbits has been investigated, it has not been adopted in cultivar development. Naturally occurring triploids have been found in pointed gourd (*Trichosanthes dioica*) and ivy gourd (*C. grandis*).

Haploids, which are highly sterile, occasionally occur spontaneously in cucumber and melon. In cucumber, they can be recognized by their reduced seed weight. Haploids can also be produced by interspecific hybridization, or pollination with irradiated pollen followed by embryo culture. Haploids treated with colchicine produce homozygous diploid lines more quickly than inbred lines can be developed by self-pollination. Doubled haploids are used extensively now to produce inbred lines quickly following a cross. These enable plant breeders to develop cultivars faster than before. Doubled haploids, or dihaploids, are considered one of the tools of speed breeding.

Polysomaty, in which the chromosome number of some somatic cells of a plant are multiples of the typical chromosome number for that plant, has been detected in melon, squash and other cucurbits. It is believed that triploid plants of pointed gourd were propagated vegetatively from triploid shoots on diploid plants (Singh, 1990).

In squash, sterile interspecific F_1 hybrids have been treated with colchicine to produce amphidiploids (allotetraploids). Self-fertile amphidiploid lines with the parentage of *Cucurbita moschata* × *C. maxima* have been produced that segregated for some horticulturally favourable characteristics. However, amphidiploids have not been important in the development of new cultivars.

Triploid interspecific hybrids of *Cucurbita* have been produced, some combining the genomes of three different species. Interspecific triploids have also been backcrossed with one of the diploid parents to create fertile interspecific trisomics. Interspecific trisomics of *C. moschata* and *C. palmata*, combining 20 chromosomes of the former and one of the latter, were synthesized and used to relate genes to specific chromosomes (Graham and Bemis, 1990). Recently, the use of trisomics to map genes on to chromosomes has been replaced with the use of molecular markers and genome sequencing.

GENES

Genetic knowledge of cucurbits is behind that of maize, tomato and pea, despite the considerable natural genetic variation in many species of the Cucurbitaceae. The use of winter greenhouses, trellises and cages has reduced the space and labour requirements for crossing cucurbits for genetic studies. The establishment of the Cucurbit Genetics Cooperative, along with the publication of the *Cucurbit Genetics Cooperative Report* annually since 1978, has fostered communication among cucurbit researchers and stimulated more cytogenetic investigations.

Dominance relationships of genes in melon were first investigated by Sagaret in the mid-19th century. Mendelian inheritance in cucumber was reported in 1913, and many genes for each of the major cucurbit crops have since been identified. In 1976, a total of 170 individual genes were known for the Cucurbitaceae. Of these, 68 were for cucumber, 37 for melon, 30 for squash species, 25 for watermelon and ten for other genera. Numerous other genetic factors were known for cultivated cucurbits and used in breeding programmes, but were not included in the gene list because their inheritance was complex or unknown.

Many additional cucurbit genes and alleles have been identified in the intervening decades and, with the application of genomics to the various cucurbit species, the pace is accelerating. The Cucurbit Genetics Cooperative publishes gene lists for the major cucurbit crops and the earlier gene lists included 146 loci for cucumber (Wehner, 1993), 100 for melon (Pitrat, 1994) and 81 for watermelon (Rhodes and Zhang, 1995). As of 2014, the number of genes published was 509, with 167 for cucumber, 160 for melon, 93 for squash species, 62 for watermelon and 27 for other genera. Thus, the number of known genes for the cucurbit crops has tripled in the past 40 years. Some of those genes code for isozymes. In addition, multiple alleles have been identified, e.g. $Y^{Scr} > Y^{Crl} > y^{O} > y$ and $G > g^{W} > g^{M} > g^{N} > g$ loci in watermelon. It should also be mentioned that identification of a gene does not necessarily indicate specific knowledge about the location, structure, or sequence of that gene. In most cases, many of these genes do not yet have molecular markers associated with them.

Regarding gene nomenclature, it should be pointed out that most genes (though not those of enzymes) are recognized and named according to the discovery of an atypical expression of that gene, which is itself caused by a newly found allele at a previously unrecognized locus. The gene and atypical, or 'mutant', allele are given the same designation (e.g. *gl* is an allelic form of gene *gl*). The first letter of the symbol is in lower case if the atypical allele is recessive and uppercase if dominant. The designation for the normal allele of that gene is given as +, the gene name with a superscripted +, or the gene name with the first letter in the case (lower or uppercase) opposite that for the atypical allele. Continuing the above example for the glabrous (*gl*) gene, the common or normal allele can be designated as +, gl^+ or Gl. In this volume, we use the superscript system for referring to common alleles.

When additional alleles are found and assigned to a gene, they are designated with different superscripts appended to the gene name. However, allelism testing runs far behind the discovery of genetic anomalies. When several alleles affect the same heritable trait, they are often treated as belonging to different genes until proven otherwise, although allelism tests are recommended before proposing another locus. This may be a problem for many alleles relating to disease resistance in cucurbits. More testing is needed to assign various alleles to their proper gene locus and to determine gene linkages.

Cucumber

Many alleles conferring disease resistance have been found and incorporated into cucumber cultivars. Dominant alleles have been reported for resistance to bacterial wilt (*Bw*), scab (*Ccu*), target leaf spot (*Cca*), Fusarium wilt (*Foc*) and watermelon mosaic virus (*Wmv* and *wmv-1-1*). Linkage of scab resistance with Fusarium wilt resistance has resulted in many cultivars having both, even though they were only selected for scab resistance. A dominant allele (*Cmv*) at a single locus has also been postulated for resistance to cucumber mosaic virus, but alleles of additional genes are needed for a high level of resistance. Recessive alleles provide resistance to angular leaf spot (*psl*), zucchini yellow fleck virus (*zyf*) and zucchini yellow mosaic virus (*zymv*). Alleles *Ar* and *cla* reportedly confer resistance to different races of anthracnose. Resistance to papaya ringspot virus is provided by *prsv* or *Prsv-2*. Several genes have been proposed to govern resistance to downy mildew and powdery mildew, with the main ones being the closely linked *dm* and *pm*.

The bitterfree loci (*bi* and *bi-2*) inhibit biosynthesis of cucurbitacin, which is an attractant for cucumber beetles but a repellant for spider mites, aphids and various other insects. The *bi* genes are epistatic to the bitter genes (*Bt* and *Bt-2*) for increased cucurbitacin content.

Salinity tolerance is influenced by many genes, in addition to a single major gene (*sa*). A single gene (*Sd*) has also been reported to control resistance to sulfur dioxide air pollution.

Dwarf plant habit, due to short internodes, is produced by alleles *bu*, *by*, *cp*, *cp-2* and *dw*. Although these alleles are currently assigned to distinct loci, allelism testing is needed to confirm that none represent allelic variants of the same gene. Allele *dw* also retards the development of oversized, and hence unmarketable, fruit. Vine size is also reduced by the determinate habit allele (*de*), which is modified by the intensifier gene *In-de*.

The interaction of two major genes, *m* and *F*, influences sex expression. *F* is modified by *In-F* and alleles at other genes. Some cultivars heterozygous for *F* have only female flowers, due to the right combination of modifying genes, but others can be monoecious under some growing conditions. Cultivars homozygous for *F* have been developed because they are more dependably gynoecious than heterozygous cultivars. The use of these genetic backgrounds in the production of gynoecious hybrid seed is described in Chapter 7. Additional genes have been reported to influence androecious (*a*), andromonoecious (*m-2*) and gynoecious (*gy*) sex expression.

Multipistillate alleles *mp* and *Mp-2* and their modifiers at other loci increase the number of pistillate flowers per node. The gene for twin fused fruit (*tf*) results in two fruit at one node fusing into a single unit. The development of parthenocarpic fruit is governed by *Pc* and modifying genes.

Fruit spine colour is governed by the putative genes *B, B-2, B-3* and *B-4*, with black spines being dominant to white. Spine number and size is influenced by genes *ns, ss, s-1, s-2* and *s-3*. Spines and warts are absent on fruit of plants with the *gl* allele for glabrous foliage and more pronounced when possessing the tuberculate fruit allele, *Tu*.

European glasshouse cultivars have glossy fruit with a tender skin and uniform dark green colour, without light green stippling. These are monogenic traits, governed by the *D* gene for dull versus glossy fruit and the tender skin (*te*) and uniform colour (*u*) genes.

Green immature fruit colour is dominant to white (*w*) and yellow green (*yg*). The interaction of alleles at two genes, *wf* and *yf*, reportedly determines white versus yellow or orange flesh colour.

Using qualitative genetics, cis linkage groups were proposed for cucumber genes (Pierce and Wehner, 1990) based on recombination frequencies. Using molecular genetics, three accessions of cucumber ('Chinese Long', Gy 14, and PI 183967) have been sequenced and the data made available on the Cucurbit Genomics Database. Chromosome assignments of the linkage groups have been made, and comparison with other species for shared synteny has shown how the cucumber chromosomes evolved (Huang *et al.*, 2009).

Melon

Many loci governing disease resistance of melon have been investigated. A single gene, *Ac*, governs Alternaria leaf blight resistance in melon line MR-1. Resistance to races 0 and 2 of Fusarium wilt is provided by *Fom-1* or *Fom-3*, and there is an allele of another gene (*Fom-2*) for resistance to races 0 and 1. Allele *Mc* confers a high level of resistance to gummy stem blight, and *Mc-2* a moderate degree of resistance to that disease. Five independent dominant genes, *Gsb-1* through *Gsb-5*, have also been reported to confer high levels of gummy stem blight resistance, from PIs 140471, 157082, 511890, 482398 and 482399, respectively. Eight loci have been postulated to influence resistance to the powdery mildew disease caused by *Podosphaera xanthii* and three more to the powdery mildew incited by *Erysiphe cichoracearum*. Four genes have been reported for downy mildew resistance. Two alleles (Prv^1 and Prv^2) of a single gene provide resistance to papaya ringspot virus from PI 180280, but they differ in reaction to some strains of that virus. Prv^2 is recessive to Prv^1 but dominant to Prv^+. Resistance to pathotype 0 of zucchini yellow mosaic virus is provided by dominant allele *Zym* from PI 424723. A single dominant gene, *Wmr*, confers partial resistance to watermelon mosaic virus. Although polygenic resistance to cucumber mosaic virus has been known, recently a single dominant gene, *Creb-2*, has been reported to confer resistance to CMV-B2 strain. Two complementary recessives genes, *cab-1* and *cab-2*, confer resistance to cucurbit aphid-borne yellows virus (CABYV). Resistance to cucurbit yellow stunting disorder virus (CYSDV) has been attributed to a single dominant gene (*Cys*) from TGR-1551. Subsequent studies have suggested that a single recessive gene may provide stronger resistance. Another dominant gene, *Liy* from PI 313970, confers resistance to the related Crinivirus, lettuce infectious yellows virus. The same PI has recessive resistance to cucurbit leaf crumple virus, conferred by a single gene, *culcrv*. Another single recessive gene, *nsv*, confers resistance to melon necrotic spot virus. Resistance to this virus is also attributed to two independent dominant genes, *Mnr-1* and *Mnr-2* (McCreight *et al.*, 1993; Lopez-Sese and Gomez-Guillamon, 2000; Frantz and Jahn, 2004).

Several melon genes for insect resistance have been identified. *Af* influences resistance to red pumpkin beetle. Tolerance to melon aphid is provided by the dominant allele at gene *Ag*, and *Vat* conditions resistance to viruses transmitted by that pest. Two complementary genes, *dc-1* and *dc-2*, govern resistance to melon fruit fly. Genes controlling foliar cucurbitacin content, including *cb* and *Bi*, influence insect resistance; plants homozygous recessive at both of these loci are resistant to cucumber beetles. A single dominant gene, *Lt*, confers antibiosis resistance to leafminer (*Liriomyza trifolii*) (Dogimont *et al.*, 1999).

Many genes influence melon plant habit, several of which have been identified. Compact habit can be obtained by breeding for the homozygous recessive state at one of the short internode loci, *si-1*, *si-2* or *si-3* (Knavel, 1990) or short lateral branching, *slb* (Fukino *et al.*, 2012). Fine mapping of short internode loci identified a region on chromosome 7 (Hwang *et al.*, 2014). Allele *Imi* increases internode length on the main stem, and *ab* (abrachiate) inhibits lateral branch development.

Genes *a* (andromonoecious) and *g* (gynomonoecious) interact to influence sex expression in the following manner: monoecy ($a^+/- \ g^+/-$), andromonoecy ($a/a \ g^+/-$), gynomonoecy ($a^+/- \ g/g$) and hermaphrodism ($a/a \ g/g$). Stable gynoecious sex expression can be achieved by combining homozygous recessive *gy* (gynoecious) at a third locus with a dominant allele at the *a* locus and the *g* allele in the homozygous state at the *g* locus (i.e. $a^+/- \ g/g \ gy/gy$). Genetic markers for the *a*, *m*, and *g* loci have been developed (Noguera *et al.*, 2005; Feng *et al.*, 2009; Gao *et al.*, 2011). Melon plants are male sterile if homozygous recessive for alleles at one of the independent genes *ms-1*, *ms-2*, *ms-3*, *ms-4* or *ms-5*.

Fruit of some melon cultivars detach (slip) from the vine at maturity due to the presence of an abscission layer, but other cultivars, lacking this trait, have persistent (non-slip) fruit. Two dominant alleles, *Al-1* and *Al-2*, control formation of the abscission layer.

Fruit quality is a polygenic trait, but several individual genes have a major effect. In wild melon populations, fruit may be bitter due to *Bi* and have mealy flesh texture because of the *Me* allele. Sour taste is dominant to sweet and conditioned by the *So* gene. A recessive allele (*if*) for juicy flesh and a dominant allele (*Mu*) for musky flavour have been reported. A recessive gene (*suc*) has been reported to condition high sucrose accumulation in melon.

Many genes influence the intensity of flesh colour, but individual major genes may determine whether the flesh is orange (which is dominant), green or white. Recessive alleles of two genes govern green flesh (*gf*) and white flesh (*wf*), with *wf*⁺ epistatic to *gf*⁺/*gf*.

External fruit colour is influenced by *Mt* (mottled rind), *st* (striped epicarp), *w* (white mature fruit), *Wi* (white immature fruit) and *Y* (yellow epicarp). Fruit shape genes include *O* (oval), *s* ('sutures' or vein tracts) and *sp* (spherical fruit shape).

A molecular linkage map of melon with 12 linkage groups, corresponding to the 12 chromosomes, has been assembled by groups in Texas, France and Spain. Additionally, the melon genome has been sequenced in Spain and is now available to further enhance the genetic map saturation (Oliver *et al.*, 2001).

Squash

Most descriptions of genes with links to the primary literature can be found in Paris and Padley (2014). In the most recent update, a number of new disease resistance genes have been added for squash. Powdery mildew resistance of *C. okeechobeensis* (Small) L. H. Bailey and *C. lundelliana* L. H. Bailey is controlled by dominant alleles of the *Pm-0* and *Pm* loci, respectively, and modifier genes may influence expression of these alleles. Two recessive resistance genes (*pm-1* and *pm-2*) have been characterized in *C. moschata*. Three dominant complementary genes (*Crr-1*, *Crr-2* and *Crr-3*) for resistance to crown rot (caused by *Phytophthora capsici*) were introgressed from *C. lundelliana* and *C. okeechobeensis* ssp. *okeechobeensis* into *C. moschata*. Six genes plus modifiers originating from *C. moschata* have been reported to affect resistance to zucchini yellow mosaic virus. For four loci, the resistant allele is dominant, whereas for the other two loci, alleles are recessive. *Zym-0* and *zym-6* appear to act independently, whereas *Zym-1*, *Zym-2* and *Zym-3*, and *Zym-4* and *zym-5* form complementary sets. A modifier of *zym-6* (*m-zym-6*) alters recessive expression to dominant expression. Recessive resistance (*zym*ᵉᶜᵘ) has also been found in *C. ecuadorensis* to ZYMV. Watermelon mosaic virus resistance has been described from two different species with *Wmv* from *C. moschata* and *Wmv*ᵉᶜᵘ identified in a *C. maxima* ×

C. ecuadorensis interspecific cross. A pair of complementary genes (*prv-1* and *Prv-2*) identified in *C. moschata* confer resistance to papaya ringspot virus. Two genes (dominant *Slc-1* and recessive *slc-2*) independently confer resistance to squash leaf curl virus in *C. pepo*. A single gene for cucumber mosaic virus resistance (*Cmv*) was identified in *C. moschata*. In some cases, such as for *Zym-1* and *Wmv*, resistance to different viruses may be conferred by genes at the same locus, or genes that are tightly linked in a complex locus. Homology among resistance genes to the same pathogen from different species has yet to be resolved.

Resistance to the melon fruit fly is provided by allele *Fr* in *C. maxima*. *C. pepo* plants homozygous for *cu*, which reduces foliar cucurbitacin content, are not preferred by cucumber beetles since these insects are attracted by cucurbitacins. Resistance to silver leaf disorder caused by whitefly (*Bemisia tabaci*) is found in both *C. pepo* and *C. moschata* with resistance conferred by recessive *sl*.

A number of genes are associated with domestication traits in *Cucurbita*. Bitterness caused by cucurbitacin is under the control of several genes. In addition to *cu* which conditions bitter cotyledons, *Bi-0* conditions bitter fruit in *C. pepo*. Three bitter fruit loci (*Bi-1*, *Bi-2* and *Bi-3*) were found in an interspecific cross of *C. pepo* × *C. argyrosperma*. *Bi-0* and either *Bi-1* or *Bi-2* may be allelic. In *C. maxima* and *C. maxima* × *C. ecuadorensis* crosses, *Bi*^{max} conditions bitterness. Wild *Cucurbita* species and some summer squash cultivars have a hard fruit rind. Rind hardness is influenced by the *Hr* (hard rind) gene of *C. pepo* and *Hi* (hard rind inhibitor) of *C. maxima*. In the presence of *Hr*, *Wt* confers wartiness in *C. pepo*. Wartiness is also found in *C. maxima* but has not been characterized genetically. Bush habit due to short internodes is conditioned by allele *Bu* in *C. pepo*; expression of *Bu* is dominant in early stages of plant development and recessive as plants vine out. Bush habit of *C. moschata* is also under monogenic control, whereas two to three incompletely dominant genes have been implicated in different bush phenotypes of *C. maxima*. There is extreme variation in bush growth habit among species, but only one gene has been definitively described and additional ones may await discovery. While most genes affecting fruit size and shape are quantitative, two qualitative genes have been described that affect shape. Disc fruit shape is governed by the *Di* gene of *C. pepo*. *Bn* conditions butternut fruit shape and is dominant to crookneck fruit shape in *C. moschata*. The characteristic post-cooking stringiness of the flesh of 'Vegetable Spaghetti' (*C. pepo*) is reportedly governed by a single gene, *fl*. A recessive allele of a major gene (*n*) and alleles at modifying genes for naked seed reduce deposition of lignin and cellulose in the schlerenchyma and subepidermal tissue of the seed coat in *C. pepo*. Resistance in *C. moschata* to the herbicide trifluralin is governed by gene *T*, which is modified by an inhibitor gene, *I-T*.

Several genes affect flower colour, morphology and sex expression. In *C. moschata*, *de* conditions determinate plants that terminate in pistillate flowers. A single gene (*G*) governing gynoecious sex expression occurs in *C. foetidissima*, but the dominant allele has not been transferred to the cultivated squash species. All staminate flowers are conditioned by recessive *a* in *C. pepo* while *ae* increases the expression of staminate flowers in the same species. *Cucurbita* inflorescences are typically solitary, but plants that are homozygous recessive for *mf* in *C. pepo* will produce multiple flowers in an inflorescence. Three male-sterile genes, *ms-1*, *ms-2* and *ms-3*, have been reported, with *ms-1* and *ms-3* present in *C. maxima* and *ms-2* in *C. pepo*. Sterile flowers conditioned by *s-1* and *s-2* have been observed in *C. maxima* and *C. pepo*, respectively. *Cr* and *I* act in concert to modify flower colour with *Cr-I-* conditioning intense yellow, *Cr-ii* pale yellow, *crcrI-* cream and *crcrii* white in colour. These genes were discovered in a *C. moschata* × *C. okeechobeensis* cross.

Fruit colour of different squash cultivars is quite diverse, and a number of fruit pigment genes have been investigated. The incompletely dominant allele of the *B* gene, (termed precocious yellow) which was found in an ornamental gourd, has been used to breed *C. pepo* cultivars with golden fruit colour and the *B* gene usually elevates the carotenoid content of mesocarp tissue. This allele affects many other horticultural traits, such as reduced storability and increased sensitivity to chilling injury. The fruit colour expression of the *B* allele is influenced by pigmentation extender genes *Ep-1* and *Ep-2*, suppressor gene *Ses-B* and other modifying genes. Fruits have a yellow and green bicolour appearance when *B* is heterozygous. This trait has been transferred to *C. moschata* but naturally occurring precocious yellow types occur within this species, and that gene is also incompletely dominant. Yellow to orange fruit colour of *C. maxima* cultivars is controlled by the B^{max} gene and modifiers. Unlike *B* in *C. pepo* and *C. moschata*, B^{max} is completely dominant. Two genes of *C. pepo*, *l-1* and *l-2*, govern light pigmentation of fruit. Allele *l-1*St (striped fruit) is recessive to allele *1-1*$^{+}$ but dominant to *l-1*. The dominant gene *D* confers dark stem colour, especially the peduncle, versus light green stems in plants homozygous for *d/d*, and is epistatic to *l-1* and *l-2*. The allele *D*^{s} darkens stems but not the fruit rind. Two genes, *Wf* for white flesh and *W* for weak pigmentation, together with *l-1* and *l-2*, govern white rind colour in *C. pepo*. The *W* gene is also epistatic to the effects of *D* on rind colour of fruit but not stem pigmentation. The *Y* allele confers development of yellow fruit shortly after anthesis in *C. pepo*. Green fruit colour in *C. moschata* is governed by gene *Gr*. The *Mldg* gene of this species determines whether immature fruits are mottled light and dark green or coloured uniformly dark green. In *C. maxima*, the *bl* allele produces blue-grey fruit skin colour in combination with genes for green skin, but pink fruit in plants carrying the B^{max} allele.

The use of molecular markers for linkage map development in squash is catching up with other cucurbit crops. Early maps were based on interspecific combinations but as marker systems and genome coverage has improved, more recent maps have been based on biparental intraspecific crosses. One of the first linkage maps for *Cucurbita* that used markers other than morphological ones was based on the cross of *C. maxima* × *C. ecuadorensis*, where five linkage groups of isozyme genes were proposed by Weeden and Robinson (1986). An interspecific backcross population of *C. pepo* × *C. moschata* was used to map 148 RAPD markers in 28 linkage groups. Loci for three qualitative genes were mapped along with two QTL (Brown and Myers, 2002). Zraidi *et al.* (2007) developed a consensus map from two *C. pepo* recombinant inbred populations using a set of RAPD, AFLP and SCAR markers. Eight QTL associated with seed coat lignification were also identified through bulked segregant analysis. Gong *et al.* (2008) created separate intraspecific maps for *C. pepo* and *C. moschata* using SSR markers. The use of microsatellite markers allowed the comparison of linkage maps of the two species to determine levels of macrosynteny. The first intraspecific *C. maxima* map was created by Ge *et al.* (2015) using a combination of SSR, AFLP and RAPD markers. A more complete interspecific *C. maxima* × *C. ecuadorensis* using RAPDs was developed (Singh *et al.*, 2011). Genotype by sequencing was applied to two species to develop high-density maps based on SNPs. Zhang *et al.* (2015) created an SNP map for *C. maxima* and used QTL analysis to identify a candidate gene for vine length. For *C. pepo*, the genome was sequenced and an SNP-based linkage map was generated in a summer squash background (Montero-Pau *et al.*, 2017).

Watermelon

Two genes of watermelon have been reported to govern anthracnose resistance, *Ar-1* for race 1 and *Ar-2-1* for race 2 resistance. Alleles *db* and *Fo-1* provide resistance to gummy stem blight and race 1 of Fusarium wilt, respectively. However, gummy stem blight resistance appears to be due to more than just a single gene. Susceptibility to powdery mildew is governed by *pm*. Resistance to race 2 powdery mildew from PI 270545 is controlled by at least two genes. Resistance to the watermelon strain of papaya ringspot virus (PRSV-W) is controlled by a single recessive gene, *prv*. A moderate level of resistance to zucchini yellow mosaic virus was conferred by a single recessive gene *zym-FL*. A high level of resistance to the Florida strain of zucchini yellow mosaic virus was controlled by a single recessive gene, *zym-FL-2*; not the same as *zym-FL*. Resistance to the China strain of zucchini yellow mosaic virus was controlled by a single recessive gene *zym-CH*. A single dominant allele, *Zym*, confers resistance to zucchini yellow mosaic virus. Insect-resistance genes for watermelon include *Af* (red pumpkin beetle resistance) and *Fwr* (fruit fly resistance).

Watermelon plants with short vines can be bred if they are homozygous recessive at the *dw-2* locus or for one of the two alleles for dwarf plant habit at the *dw-1* locus. Both genes make a super dwarf that has been used to develop cultivars for patio and container production. Branching at the lower nodes of the main stem is reduced by allele *bl*.

Andromonoecy is recessive to monoecy and conditioned by gene *a*. Two alleles are known that produce male sterility, *ms* and *gms*, the latter associated with glabrous foliage.

When plants are homozygous recessive for the *e* (explosive rind) allele, the fruit is tender (not tough rind, but can be thin or thick rind), bursting when cut. The *f* gene determines furrowed fruit surface. The incompletely dominant allele *O* governs elongate versus spherical fruit shape. The *su* allele suppresses fruit bitterness.

Dark green skin colour (*G-1* and *G-2*) is dominant and other alleles at the *g-1* locus determine the degree of colour and striping: *G* (medium or dark solid green); g^W (wide stripe); g^M (medium stripe); g^N (narrow stripe); and *g* (solid light green or grey). The dominance series is $G > g^W > g^M > g^N > g$. Greenish mottling of the exocarp is produced by the *m* allele, and pencilled lines by *p*. Golden mature fruit colour and chlorosis of older leaves is governed by gene *go*.

A gene with a dominant allele for white flesh (*Wf*) is epistatic to a gene for yellow flesh; the double recessive is red-fleshed. Allele *C* produces canary yellow flesh colour. The darkness of red colour in the flesh (and the amount of lycopene) is controlled by multiple alleles at the *y* locus. Scarlet red flesh (Y^{Scr}) is dominant to coral red flesh (Y^{Crl}), orange flesh (y^O) and salmon yellow flesh (*y*). The dominance series is $Y^{Scr} > Y^{Crl} > y^O > y$. The allele Y^{Scr} is from 'Dixielee' and 'Red-N-Sweet'; the allele Y^{Crl} is from 'Angeleno' (black-seeded); the allele y^O is from 'Tendersweet Orange Flesh'; and the allele *y* is from 'Golden Honey'.

The interaction of alleles at several genes, including *d* (dotted seed coat), *r* (red), *t* (tan) and *w* (white seed coat), determine seed coat colour and pattern. Clump is *RR TT ww*; tan is *RR tt WW*; white is *RR tt ww*; green is *rr TT WW*; red is *rr tt WW*; and white with pink tip is *rr tt ww*. Seed size is controlled by several genes, including *l* (long) and *s* (short seed) genes, with *s* epistatic to *l*, and long recessive to medium or short. The phenotypes are *LL SS* for medium, *ll SS* for long, and *LL ss* or *ll ss* for short seed. The sources for genotypes are *ll SS* from 'Peerless', *LL SS* from 'Klondike', and *LL ss* from 'Baby Delight'. A third gene, *ts*, produces tomato seed size, which is smaller than short seed. Also, tiny seed (*Ti*, dominant to short seed) from 'Sweet Princess' produces seeds that are between short and tomato seed sizes. Five genes for seed protein composition (*Spr-1, Spr-2, Spr-3, Spr-4* and *Spr-5*) are included in the watermelon gene list.

Four linkage groups of 13 isozyme genes were reported by Navot and Zamir (1986). Since then, the *Citrullus* genome has been sequenced using the Chinese breeding line 97103 (Guo *et al.*, 2013), and later using 'Charleston Gray'.

GENOMICS

With the advent of next-generation sequencing technologies, genomes of some of the major cucurbits have been sequenced, creating new sets of resources for these crops. Genomes of watermelon, melon, cucumber, squash (*C. pepo* and *C. moschata*), pumpkin (*C. maxima*), and bottle gourd (*L. siceraria*) are now available, and the list will continue to grow as sequencing costs decrease (see Table 3.1). The information is available on the Cucurbit Genomics Database maintained by J. Fei at Boyce Thompson Institute (Cornell University, Ithaca, New York).

Cucumber, the first cucurbit genome to be sequenced (2009), enabled scientists, including plant breeders, to identify novel biosynthetic pathways and to develop molecular markers. With each cucurbit species genome that has been sequenced, we have obtained a greater understanding of the evolution of the Cucurbitaceae as well as the function of genes shared across species. The high synteny between genetic sequences of cucumber and melon showed the chromosome fusions that led to cucumber from melon. The bottle gourd genome, one of the most recent species sequenced, brought new evidence that ancient cucurbits had 12 chromosomes. While most major cucurbit crops have had whole-genome sequencing completed, many minor cucurbits have had partial genome sequencing performed. Through these studies, novel and species-specific genes have been identified for bitter gourd (Behera *et al.*, 2016), wax gourd (Jiang *et al.*, 2013), wild *Cucumis* species (Ling *et al.*, 2017) and other cucurbits, including the creeping perennial, *Gynostemma pentaphyllum* (Chen *et al.*, 2016).

These resources have been leveraged for dozens of studies on transcriptome analyses of developing fruit, disease resistance or susceptibility, growth under abiotic stress conditions, and many other stages of plant growth in cultivated and wild cucurbits. Many transcriptome studies to date in cucurbits have focused on cucumber because of its economic importance (Ando *et al.*, 2012; Kong *et al.*, 2015; Wyatt *et al.*, 2016; Li *et al.*, 2017; Sun *et al.*, 2017). While few cucurbit transcriptome studies have resulted in the identification of a single gene controlling a specific trait, these large-scale transcriptome studies provide valuable information on genetic pathways, patterns of expression, and candidates for breeding and genetic transformation technologies.

HISTORICAL BREEDING TECHNIQUES

The nature of DNA in cucurbits has been under study for more than three decades. Bhave *et al.* (1986) analysed the distribution of repeat and single copy DNA sequences in smooth and angled luffa, wax gourd and ivy gourd. Around the same time, Ganal and Hemleben (1986) compared restriction enzyme maps of ribosomal DNA repeats in cucumber, melon, and two squash species, *C. maxima* and *C. pepo*. An early and continuing objective for studying DNA and RNA in cucurbits has been to clarify phylogenetic relationships within the family. Analysis of variation of chloroplast DNA has revealed species relationships in *Cucurbita* (Wilson *et al.*, 1992) and *Cucumis* (Perl-Treves and Galun, 1985), and researchers continue to use DNA-based information systems to detect origins and evaluate relationships in the Cucurbitaceae (Schaefer *et al.*, 2009; Chomicki and Renner, 2014).

Speed breeding is a system that makes use of new technology to develop cultivars in less time. Some of those technologies that have been applied to speed breeding of cucurbits are as follows. Greenhouses can be supplied with lights that permit faster growth and flowering in the winter season. In addition, temperature control permits rapid growth, usually with the objective of 32°C day and 21°C night. Field trials can be run faster by using many locations and few years. Speed breeding usually builds on previous technologies. Some of those include winter nurseries, optimum size of roots in greenhouse pots or bags, and the use of growth regulators to increase the speed of making crosses and self-pollinations.

Much of the early progress made towards understanding the genetic code of cucurbits and other plants was due to the use of restriction enzymes, also called endonucleases. An endonuclease cleaves DNA at a particular location based on the enzyme's recognition of a certain sequence of nucleotides. There are many different endonucleases, each cutting DNA at different places, and creating DNA fragments of various lengths. Endonucleases allowed for DNA mapping, which has been going on for cucumber, melon and squash for the past 30 years. For example, Gounaris *et al.* (1990) used endonucleases to resolve the genetic map of chloroplast DNA in *C. pepo*, and more recently, endonucleases were used to generate a genetic map of zucchini (*C. pepo*) using genotyping by sequencing (GBS) (Montero-Pau *et al.*, 2017).

Endonucleases also enable scientists to create recombinant DNA, which results when cleaved DNA strands from two different DNA molecules are enzymatically joined. If one of these molecules is a bacterial plasmid (an independent, self-replicating ring of DNA in the bacterial cell), then DNA from the other molecule can be incorporated into the plasmid. The plasmid containing the recombinant DNA can be transferred to embryonic plant cells, which, in turn, pass the genetic material to new cells.

When the transgenic vector is a bacteriophage (a bacteria-infecting virus) instead of a plasmid, the recombinant DNA is passed from virus to bacterium (where it is incorporated into the bacterial DNA), and then from bacterium to plant. Bacteria containing incorporated viral DNA possess a degree of immunity to attacks by the same phage because the incorporated DNA directs synthesis of a phage repressor molecule that regulates the expression of the viral DNA.

NOVEL BREEDING TECHNIQUES

Speed breeding has advanced as new technologies have emerged. Doubled haploid production has made it faster to produce inbred lines, although it is sometimes difficult to produce enough dihaploids for each cross that plant breeders are interested in evaluating. Often, hybrids are produced before the parental lines have been thoroughly evaluated. In that way, the best hybrids are quickly available when the data from field and greenhouse trials have identified those that should be advanced to the next stage of breeding. In the same way, lines are used to start new crosses before they have been thoroughly evaluated. Seedling evaluations using molecular markers to select individuals with traits of interest are also decreasing generation times.

Numerous systems to detect variability in plant DNA have been developed and used to facilitate genetic mapping. Simple sequence repeats (SSRs) and single nucleotide polymorphisms (SNPs) are two of the most common molecular marker systems used for generating genetic maps currently. One use of genetic mapping is for the identification of molecular markers associated with traits of interest (high yield, fruit quality, disease resistance). Linkage between a molecular marker and a trait of interest allows one to identify the presence of the trait by checking the genotype at the marker locus. This testing can occur as early as the seed stage, allowing for rapid selection of plants that have the trait of interest. In cucumber, watermelon, squash and melon, markers are available to select for improved fruit quality and disease resistance.

In cases where simply inherited sources of disease resistance have not been found (for example, resistance to CMV in melon and squash), genetic engineering and transfer of the coat protein or replicase genes of the virus to cucurbits has given these cucurbits disease resistance. For example, Chee and Slightom (1991) developed the coat protein-mediated mechanism for CMV protection in cucumber. They used a disarmed strain of *Agrobacterium tumefaciens* to transfer the engineered viral coat protein to cucumber. Small pieces of young, growing cotyledon tissue were soaked in a solution containing the recombinant DNA-carrying bacterial plasmids. The transformed embryogenic calli were regenerated and the resulting plants were inoculated and shown to have resistance to CMV. Gonsalves *et al.* (1992) went on to prove that the transgenic plants had a high level of CMV resistance under natural field conditions, where viral exposure is usually via insect vectors.

Similar field performance experiments have been conducted for transgenic squash lines (*C. pepo*). The virus-susceptible hybrid 'Pavo' was compared with two transgenic selections from this cultivar that possessed CMV and WMV resistance (Arce-Ochoa *et al.*, 1995). Under field conditions in Texas, the two transgenic selections had only 3% and 14% symptomatic plants, compared with 53% for 'Pavo'.

In the USA, the first commercially available transgenic cucurbit was 'Freedom II' (previously called 'ZW-20'), a yellow summer squash of the crookneck type (*C. pepo*). This cultivar was engineered with resistance to WMV and ZYMV. Again, *A. tumefaciens* was used, this time to transfer the coat protein genes of WMV and ZYMV to crookneck squash. Since the introduction of 'Freedom II' additional cultivars have been released, including 'Liberator III' and 'Independence II'.

More recently, gene editing techniques have gained popularity. The clustered regularly interspersed short palindromic repeats (CRISPR)-associated protein system is universally recognized as a powerful genome editing tool. In brief, CRISPR is an RNA-guided editing tool that uses a short (21–72 nucleotides) seed sequence flanked by palindromic repeats to recognize and cleave a specific sequence of DNA. Unlike genetic engineering, where genes are inserted at random, genome editing allowed for specific changes with no to few off-target effects or foreign genes remaining in the new cultivar. CRISPR represents a highly efficient, low-cost genome editing tool and is now being developed for cucurbits. While the CAS9 system is the most common, multiple CRISPR systems are being developed to fully utilize this technology. In cucumber, CRISPR has been used to incorporate resistance to zucchini yellow mosaic virus, papaya ringspot mosaic virus-W and cucumber vein yellowing virus into a non-transgenic plant (Chandrasekaran *et al.*, 2016). In watermelon, a CRISPR system has been tested using phytoene desaturase, but new quality traits have not yet been successfully incorporated (Tian *et al.*, 2017).

Major Crops

CUCUMBER, MELON AND THEIR RELATIVES

Introduction

A biosystematic monograph of the genus *Cucumis* (Kirkbride, 1993) recognized 32 species, including two major crops, cucumber (*C. sativus*) and melon (*C. melo*), and two minor crops, West Indian gherkin (*C. anguria*) and African horned melon (*C. metuliferus*). Other species are sometimes cultivated (e.g. *C. dipsaceus*) or collected wild and used for food, water, or medicine (e.g. *C. africanus* L.). Molecular phylogenetic studies have placed additional genera into *Cucumis* based on chloroplast and nuclear DNA sequence similarities. A recent analysis of 100 *Cucumis* species from Africa to Australasia (Sebastian *et al.*, 2010) provided a better picture of species diversity and relatedness and added 18 close relatives of cucumber and melon not previously included by Kirkbride. Of the 66 *Cucumis* species, 25 are Asian and Australian and 41 are African.

    Cucumis is divided into subgenus *Cucumis*, composed of *C. sativus* and *C. hystrix* Chakravarty, and subgenus *Melo* (Mill.) C. Jeffrey, containing the remaining species. Phylogenetic investigations of *Cucumis* species have been based on crossing relationships, karyotypes, flavonoids, isozymes, chloroplast DNA and molecular markers. All of these studies agree that the subgenera are widely separated, to the point that it has been proposed they be in distinct genera. Puchalski and Robinson (1990) proposed seven groups of *Cucumis* species according to isozyme patterns. The groupings were generally similar to those above, except that the anguria group was separated into three isozyme groups and *C. sagittatus* was placed in a different isozyme group than *C. melo*.

    Interspecific crossing relationships (see Chapter 3, Fig. 3.2) (Deakin *et al.*, 1971) and molecular phylogenies (Garcia-Mas *et al.*, 2004) suggest four groups of *Cucumis* species. Crossing relationship groups are: (i) the anguria group, composed of *C. anguria* and seven other intercompatible species (marked '2' on Fig. 3.2) with softly spiny fruit; (ii) *C. metuliferus*; (iii) *C. sativus*, including the fully compatible *C. sativus* var. *hardwickii* (Royle) Gabaev; and (iv) the melo group, which includes *C. melo, C. humifructus* and *C. sagittatus* Peyritsch in Wawra & Peyritsch. Species of the melo group have hairy immature fruit, but, unlike other *Cucumis* species, their mature fruit lack spines. Interspecific hybrids have not been obtained in this group, although fruit with partly developed seeds were obtained in the cross *C. melo* × *C. sagittatus*.

CUCUMBER

Botany

Cucumber, *Cucumis sativus*, is the fourth most important vegetable crop in the world and the most important cucurbit. Cucumbers, like most cucurbits, are indeterminate, vining (1–3 m), frost-sensitive annual plants that produce cylindrical or round fruit. Determinate cultivars, generated from a naturally occurring mutation for gibberellin sensitivity, were first introduced with 'Midget' in 1940 and have been used in home gardens and patio containers.

Usually, a single unbranched tendril develops at each leaf axil, although a tendril-less (*tl*) mutant has been described. The stems, leaves and young fruit of the plants are covered in stiff, multicellular, unbranched trichomes (Zhao *et al.*, 2015). The leaves of this species are triangularly ovate, with three to five lobes. The first cucumber cultivars were monoecious, but gynoecious, androecious, hermaphroditic, gynomonoecious, andromonoecious and trimonoecious forms are also present in this species. Current cultivars are gynoecious or predominantly gynoecious, except for monoecious garden cucumbers that produce fewer fruit per harvest, but more harvests.

Little variation in petal colour or shape has been observed in cucumber. Flowers are yellow with five or six basally fused petals (Fig. 4.1). Plants often produce male flowers initially (first three nodes) followed by alternating male and female flowers and finishing with mostly female flowers. Flowers remain

Fig. 4.1. Cucumber flower (left) and melon flower (right). Notice the size variation, and basal fusing of the six petals in the cucumber flower.

open for a single day and are pollinated by bees and other insects. Immature fruit are green at the edible stage, except in a few cultivars, where they are white or yellow. Fruit are round to oblong or narrowly cylindrical, with small tubercles (warts) and spines of trichome origin on the rind. Spine colour is associated with mature fruit colour and fruit netting. Fruit of white-spined cultivars are greenish-white to yellow at maturity and not netted. Black-spined fruit become orange or red (brown) when mature and may be netted. Fruit flesh is crisp and usually white, but is pale orange in a few cultivars. Seeds are small, white and flat.

Origin and history

Cucumber is of Asiatic origin along with the closely related wild *Cucumis sativus* var. *hardwickii*, which was first found in the Himalayan foothills of Nepal. Plants of this botanical variety are highly branched, daylength sensitive and prodigious producers of bitter fruit. The most closely related species to cucumber ($2n = 2 x = 14$) is *Cucumis hystrix* ($2n = 2x = 24$). Both species are more closely related to the species of Australia and New Guinea, such as *C. umbellatus*, than to the African species, such as *C. hirsutus*, *C. metuliferus*, *C. myriocarpus*, *C. anguria* and *C. sagittatus* (Sebastian *et al.*, 2010). Cucumber diverged from *C. hystrix* 4.6 million years ago, and the cucumber–melon relative diverged from its African relative 11.9 million years ago. Based on the recent sequencing of the *Lagenaria* genome, *Lagenaria* diverged from *Citrullus* 10.4–14.6 million years ago, from *Cucumis* 17.3–24.3 million years ago and from *Momordica* 29.2–41.0 million years ago.

The remains of *Cucumis* crops (melon or cucumber) in eastern Iran have been dated to the third millennium BC. Cucumber cultivation goes back perhaps 3000 years in India and 2000 years in China. China is considered a secondary centre of genetic diversification. Today, cucumber is one of the most important vegetable crops in that country, second only to Chinese cabbage in the total area that is cultivated.

Cucumber was probably not known to the ancient Egyptians, Greeks or Romans. The Latin *cucumis*, Greek *sikyos* and Hebrew *qishu'im* refers to the snake melon, *Cucumis melo* var. *flexuosus* (Paris, 2012). In English translations of the Bible, cucumber and melon are referred to in Numbers 11:5, but that probably should have been translated as snake melon and watermelon. Snake melon fruit can be distinguished by small hairs (fuzz) on the surface, whereas cucumber fruit are smooth, with ridges, warts, or spines.

Early travellers brought cucumber to Mediterranean countries from India through Iran, Iraq and Turkey in the 6th or 7th centuries (Paris, 2012). Cucumber probably reached Spain in the 9th century and Tunisia in the 10th century. In the early 14th century, cucumber plants were cultivated in England. There, the fruit were known as 'cowcumbers'. Portuguese explorers subsequently carried cucumber to West Africa. Columbus introduced the species to the New World, planting it in Haiti in 1494. Today, cucumber is grown throughout the world, in containers on patios in urban areas, in gardens often using trellises, on large commercial farms, in unheated high tunnels and in heated greenhouses (glasshouses or poly houses).

Uses

There are several non-food uses of cucumbers. White-skinned cultivars were grown in France in the 19th century for the production of cosmetics. Today, cucumbers are used in health and beauty products, including perfumes, lotions, soaps and shampoos. Indigenous practitioners create medical concoctions from the roots, leaves, stems and seeds.

Of course, the common use of cucumbers is as food, where it is classified as a warm-season vegetable. Cucumbers are most often consumed fresh or pickled. In China, India, Indonesia and Malaysia, they may be cooked before being eaten. The fruit are used in curries and chutney in India. Mature cucumber seeds are eaten, particularly in Asia, and the seeds can be crushed to produce an edible oil that is sometimes used in French cuisine. Young leaves and stems are cooked in southeastern Asia.

Cucumber cultivars are classified as slicers (served fresh in salads), trellis cucumbers (longer than 230 mm, so must be trained vertically to prevent fruit from curving), greenhouse cucumbers (similar to trellis, but parthenocarpic for fruit set without pollination in the greenhouse), or picklers (38–154 mm long, processed by canning after fermenting or pasteurizing) (Fig. 4.2). There are also Beit Alpha (middle eastern) type cucumbers that are the size of pickling type, but have smooth (wartless) skin and soft fruit for fresh use. They are also referred to as Persian cucumbers. In some areas, fruit of pickling cultivars are used like slicers in salads. Small-fruited pickling cucumbers are called gherkins in various countries, including India. The word 'gherkin' has been traced through medieval Latin, Arabic and Persian back to Urdu and Hindi, languages of the Indian subcontinent, the ancestral home of the cucumber (Paris *et al.*, 2011). Pickling cucumbers are also sold as pasteurized and acidified rather than fermented. Although not technically pickles, the former are increasingly preferred by consumers (two-thirds of the US market).

Generally, picklers have shorter fruit with more prominent warts than slicers. The length to diameter ratio, usually about 3:1, is important for pickling cucumbers. Most cultivars have white-spined fruit, but older cultivars may have either white or black spines. Pickling cultivars with white spines became the standard because their fruit retain green colour longer and turn cream or yellow rather than orange or red when they are mature (having hard seeds) (Fig. 4.2).

After harvest in commercial operations, slicers are usually graded, washed, cooled and waxed before being marketed (Fig. 4.3). Greenhouse cucumbers have thin-skinned fruit and so are wrapped in clear plastic before marketing. Processing-type cucumbers are prepared for overnight pickles (unpasteurized

Fig. 4.2. Mature fruit colour (red, orange, yellow, and white) in cucumber.

Fig. 4.3. Dark green and long slicers, medium length Beit Alpha, long pickling, and short pickling cucumber types.

refrigerated dills) and for brined and fermented products (see Chapter 7 and Fig. 7.5 for more details).

Over 70% of the USA cucumber crop is pickled. Yield of pickling cucumbers had increased more than threefold in the USA, from 3.61 t ha^{-1} in 1930 to 11.61 t ha^{-1} in 1990, but yield progress has slowed since then.

Cultivars and breeding

Glasshouse (greenhouse) cucumber cultivation is important in northern Europe, Asia, the Middle East and other areas. Glasshouse production of cucumber in the USA has declined in recent years, due to competition in winter from plants grown outdoors in screen tunnels in Florida, Mexico and other warm regions. Slicing-type cucumber cultivars are most commonly grown in glasshouses. Pickling cultivars have been grown in glasshouses in The Netherlands, but this practice is declining due to cost. Parthenocarpic pickling types are now available for outdoor production (areas must be free of honey bees and sources of cucumber pollen). The cultural practices associated with glasshouse cucumbers are described in Chapter 6.

Glasshouse cucumbers have become popular in markets, commanding a premium price because of their excellent quality. They can often be identified by their very long, slender fruit, with constricted neck, thin skin, indistinct warts and spines and crisp texture. The fruit are seedless and are produced on plants that are bitterfree, making the fruit mild-tasting. Their crispness, tender skin and other quality attributes are primarily due to the cultivar, rather than to being grown in a glasshouse. Glasshouse cultivars include the original European gynoecious cultivar 'Telegraph' and the related, improved selections 'Petita F$_1$' and 'Superator'. The small-fruited versions such as 'Hayat' are the new type for the Middle East, the greenhouse Beit Alpha (Bet Alfa, derived from the landrace from Damascus) or Persian cucumber. Recently, greenhouse Beit Alpha (parthenocarpic) type was developed to produce a fruit similar in size to Beit Alpha, but in high tunnels (unheated) or greenhouses.

Some of today's cultivars are known to be centuries old, having originally been developed in Europe or Asia. An example is 'Early Russian', which was described by Charles Naudin in France in 1859. Many field-grown cultivars in Europe and the Middle East differ from American cultivars by having numerous fine spines and indistinct warts.

Cultivar selection in the USA began in the late 1880s, with emphasis placed on fruit shape and colour as well as adaptation to local growing conditions. Many cultivars introduced before 1900 were developed simply by selecting superior plants from the heterogeneous cultivars grown at that time. American and English cultivars were crossed to develop 'Tailby's Hybrid', introduced in 1872, and controlled hybridization has subsequently become important in cucumber improvement. Some of the old cultivars that are still available today include pickling-type 'Early Cluster' (1778), 'Early Russian' (1854), 'Chicago Pickling' (1897), 'Snows Pickling' (1905), 'National Pickling' (1924), 'Double Yield' (1924), 'Producer' (1945), and 'Model' (1946). Slicing-type cultivars include 'Early White Spine' (1806), 'Arlington White Spine' (1886), 'Klondike' (1902), 'Davis Perfect' (1906), 'Longfellow' (1927), 'Straight Eight' (1935), 'Delcrow' (1936) and 'Marketer' (1942).

Old cultivars of the American slicing type develop stippling at high temperature (light green spots at the location of the lenticels on the fruit). Cultivars from Asia and most European glasshouse cucumbers have an allele for uniform green fruit colour (*u*), and H.M. Munger backcrossed this allele into 'Marketmore 70' and other breeding lines to produce cucumbers that retain their uniform, dark green fruit colour even at high temperatures. Most slicing cultivars marketed in the USA today have that trait.

'Maine No. 2' (released 1939) was the first cucumber cultivar resistant to scab disease. Its resistance is due to a single dominant allele. Scab is no longer an important disease in most cucumber crops, because most current cultivars have that allele, which is linked to resistance to Fusarium wilt, another important disease. W.C. Barnes developed 'Pixie' (released 1963) with resistance to diseases in production areas in the southern USA: downy mildew, powdery mildew and anthracnose. Resistance permitted a fall crop to be grown, expanding the useful production season. Later, 'Sumter' (1973) was released with additional disease resistances to angular leafspot, scab and cucumber mosaic virus (CMV).

Cultivars resistant to CMV have been available for more than 50 years. J.C. Walker combined scab and CMV resistance in 'Wisconsin SMR 18' (released 1958) pickling cucumber. A higher level of resistance to CMV, combined with good horticultural type, was achieved in the 'Marketmore' series of slicing cultivars bred by H.M. Munger. He incorporated resistance to scab and additional diseases into this cultivar by backcrossing. Cucumber breeders continue to combine genes for resistance to different diseases. C.E. Peterson developed a breeding line, WI 2757 (released 1982), with resistance to nine diseases.

'Burpee Hybrid', the first F_1 monoecious hybrid cucumber cultivar (released 1945), was bred by Oved Shifriss at the Burpee Seed Company. The development of hybrid cultivars became important after gynoecious sex expression was obtained from a Korean cultivar. The genes for gynoecious expression (*FF MM* genotype) are dominant and gynoecious hybrid cultivars have a high proportion of female flowers, resulting in early maturity and concentrated yield. Gynoecious hybrids with early and concentrated fruit set are suited to mechanical harvest. 'Spartan Dawn', introduced in 1962, was the first gynoecious hybrid cultivar. Peterson developed the maternal parent of this hybrid by backcrossing the gynoecious trait (*FF MM* genotype) into 'Wisconsin SMR 18' (*ff MM* genotype). The most popular pickling cultivars today are gynoecious hybrids.

'Lemon' is a unique cultivar. It has andromonoecious sex expression, whereas other cucumber cultivars are monoecious or gynoecious. The small, round, yellow fruit, which has five placentae instead of the customary three, faintly resembles a lemon fruit, hence its name. Other improvements made by plant breeding include high yield, fruit with high vitamin A, fruit with small seedcell, fruit resistant to damage while in brine tanks (bloater resistance), concentrated fruit set for machine harvest (using gynoecious and multiple

branching types), fruit with long length–diameter ratio for stressful environments, and parthenocarpic seedless cultivars (usually made as hybrids from
two parthenocarpic and gynoecious inbred lines) for open field production,
thus eliminating the need for pollenizer plants or pollinating insects such as
bees.

Cucumber breeders in China have developed improved inbred and F_1 hybrid
cultivars (e.g. 'Ningqin', 'JingYan No. 2') of the oriental trellis type for production
in open field as well as greenhouse (Cui and Zhang, 1991). Traits introduced
include gynoecy, earliness and multiple disease resistance. Breeders are also
trying to improve various traits for autumn crop production. Parthenocarpy
has been incorporated to reduce the need for pollination in the greenhouse. For
oriental trellis cucumbers, the preferred type is long fruit with dark green skin
and thick, prominent spines or ridges.

Breeders worldwide continue to select for a wide range of desirable characteristics, in addition to disease and arthropod (insect and spider mite) resistance. Cultivars have been bred for tolerance to cold, heat, drought, herbicides,
sulfur dioxide and soil salinity. Pickling cucumbers have been bred to withstand carpel separation in order to prevent bloating during the brining process.
Cucumber cultivars differ in the time required for their fruit to develop from
the optimal size for market to oversized fruit having little value. 'Marketer' was
an important cultivar in that respect because it produced a large proportion of
fruit of marketable stage (Munger, 1992).

The little-leaf mutant (*ll ll* genotype) was discovered in 1978 by J.L. Bowers
and M.J. Goode. Its multiple branching and multiple fruit-setting traits were
desirable for once-over harvesting systems, but there were problems with slow
fruit development (common to genotypes having multiple fruit set) that led to
fruit having tough skin, large seeds and a watery (mature) seedcell. Currently,
parthenocarpic (seedless fruit development without pollination) cultivars that
promise high yield are being evaluated for machine-harvest systems of pickling cucumbers in the USA, following their success in Europe. Problems that
need to be solved are unreliable fruit set and tough skin in the larger fruit sizes.

Genetic diversity

In cultivated cucumber and closely related species, genetic diversity is low even
though large variation in skin colour, spines and striping has been described.
While the progenitors of cucumber likely originated in Africa, initial domestication occurred in Southern Asia about 3000 years ago. Cucumber and wild
relatives from East Asia and India all have greater genetic diversity and are
genetically distinct from cucumbers developed in the USA, Europe and West/
Central China. Because of low genetic diversity present in cultivated cucumber,
related species have been studied as potential sources for new traits by breeders.
C. melo, *C. hystrix*, *C. metuliferus* and *Lagenaria siceraria* may serve as additional

sources for breeding in commercially important traits such as disease resistance. So far, only _C. hystrix_ and the closely related botanical varieties _C. sativus_ var. _harwickii_, _C. sativus_ var. _xishuangbannanensis_ and _C. sativus_ var. _sikkimensis_ have been successfully used as sources of new genes for cucumber.

Closely related species

Cucumber belongs to the genus _Cucumis_, along with other members including _C. melo_, _C. anguria_ var. _anguria_ and _C. metuliferus_, which will be discussed in detail later in this chapter. _Cucumis sativus_ var. _sativus_ belongs to the subgenus _Cucumis sativus_ along with the closely related botanical varieties _C. sativus_ var. _harwickii_, _C. sativus_ var. _xishuangbannanensis_ and _C. sativus_ var. _sikkimensis_. _C. sativus_ var. _hardwickii_ (formerly _C. hardwickii_) was initially discovered growing wild in the Himalayan foothills of India (Deakin _et al._, 1971). _C. sativus_ var. _sikkimensis_ has been reclassified as _C. sativus_ var. _sativus_. Only a few genes (such as carpel separation of the developing fruit, _Es-1 Es-1 Es-2 Es-2_, and heavy netting of the skin of mature fruit, _H H_) distinguish _C. sativus_ var. _sikkimensis_ from other cucumbers.

Crosses have been attempted between cultivated cucumber (_C. sativus_) and relatives outside its subgenus (_Cucumis_ spp.) but these have not been successful or repeatable. This, in part, is attributed to differences in chromosome number among species. The botanical varieties of _C. sativus_ are diploids with a chromosome number of $2n = 14$, while that number is usually $2n = 24$ among the other _Cucumis_ species.

The only reported successful cross with _C. sativus_ was with wild _C. hystrix_ from Yunnan Province of Southern China, a member of the secondary gene pool of _C. sativus_. In appearance, _C. hystrix_ is less appetizing than cultivated _C. sativus_ with its muricate (studded with short spikes) and verrucose fruit with black spines. It was successfully crossed with cucumber, but the resulting progeny had poor fertility and low seed set (Chen _et al._, 1997). Doubling of the chromosome number of the progeny resulted in fertile progeny of the synthetic species (amphidiploid or allotetraploid) _C. hytivus_ (Chen _et al._, 1997; Sebastian _et al._, 2010). The development of the fully fertile _C. hytivus_–derived fertile diploids ($2n = 2x = 14$) resulted in new sources of traits for cucumber improvement. Lines have been developed having vines with disease resistance and fruit with high quality. Recently, gummy stem blight resistance was mapped in three of 52 introgression lines obtained by crossing cucumber with the interspecific _C. hytivus_.

Cucumis sativus _var._ hardwickii _(wild cucumber)_
Cucumber has been improved by using accessions of the wild var. _hardwickii_ (Lower and Nienhuis, 1990). Traits such as resistance to nematodes _Meloidogyne arenaria_ and _M. javanica_ have been transferred from LJ 90430, to produce the cultivars 'Lucia, 'Manteo' and 'Shelby'.

Some traits, including little-leaf (*ll-2*), black spine (*B-3, B-4*) and multiple-branching, are not the same (are non-allelic) compared with Arkansas Littleleaf and black-spined (*B-1*) Wisconsin SMR 18, and represent new genes for those traits.

Cucumis sativus *var.* sikkimensis *(Sikkim cucumber)*

Cultivars of *C. sativus* var. *sikkimensis* Hook. f. (Sikkim cucumber) can have large, oblong fruit. The taxon has been reclassified as *Cucumis sativus* var. *sativus*. Cultivars of Sikkim cucumber are found primarily in mountainous Nepal and India. The accessions are useful in that they represent much variation, including five-carpellate fruit.

Sikkim cucumber has fruit that may be hollow and are brown and netted at maturity (*CC RR* genotype, known as red mature fruit colour). The fruit may be round or elongate, depending on accession, and will store for a few months. They may be eaten fresh, or fried like summer squash. The vines are resistant to disease, especially downy mildew, so there has been a lot of interest in these accessions recently.

Cucumis sativus *var.* xishuangbannanensis *(Xishuangbanna gourd)*

A distinct group of cultivars, known as Xishuangbanna gourd, is classified as *C. sativus* var. *xishuangbannanensis* Qi & Yuan. Grown by the Hani people of southwestern China at elevations of 1000 m or higher, these landraces are largely unknown outside eastern Asia. The Xishuangbanna gourd has much variation and includes cultivars with five-carpellate fruit. It is more closely related to *C. sativus* var. *sativus* than to *C. s.* var. *hardwickii*.

Vines of this taxon are vigorous, reaching up to 7 m in length. The large mature fruit is oblong and weighs about 3 kg. The rind is white, yellow or brownish orange, sometimes with distinct netting, but otherwise relatively smooth and without prominent warts. Because the yellow to orange flesh colour is produced by provitamin A carotenes, there has been interest for development of high-nutrition cucumbers. Researchers (e.g. Bo *et al.*, 2012) have transferred the trait (light orange mesocarp, orange seedcell) controlled by the *ore* gene from the Xishuangbanna gourd to cucumber cultivars.

Cucumis hystrix

C. hystrix is a wild species of *Cucumis* that can be crossed with *C. sativus* to produce, after chromosome doubling, an allotetraploid. The new species is *C. hytivus* and lines have been developed that have characteristics of processing or fresh-market cucumbers, along with other traits not present in cucumber. Introgression lines were produced by Chen *et al.* (2003) that have been used to introduce disease resistance to cucumber.

Resistances obtained from crosses with *C. hystrix* include gummy stem blight and downy mildew. *C. hystrix* offers a way to increase the genetic diversity of cucumber, already known to be rather narrow. No other species has been crossed successfully with cucumber, although the two botanical varieties *C. sativus* var. *hardwickii* (formerly *C. hardwickii*) and *C. sativus* var. *xishuangbannesis* will cross readily with cucumber.

MELON

Nomenclature

Many common names have been used for this species, including muskmelon, cantaloupe, nutmeg melon, winter melon (also a name for *Benincasa hispida*), sweet melon, rock melon, charentais, Italo-American, galia, piel de sapo, yellow canary and snap melon. Munger and Robinson (1991) recommend that it simply be called melon, although agricultural statistics often combine cantaloupe and watermelon into the category of 'melon', creating some ambiguity in terminology.

Botany

Melon stems are nearly round as opposed to the angular stems of cucumber, and the pubescence of melon is softer than that of cucumber. Plants typically are trailing vines, but compact cultivars with short internodes and lateral branches have been bred. Tendrils are unbranched and borne singly at the nodes. Leaves are usually subcordate and less lobed than those of cucumber. Most western shipper and honeydew melon cultivars are andromonoecious and have round or oval fruit, but some eastern shippers, 'Harper' types, 'Banana' and snake melons are monoecious with pistillate flowers that produce elongate fruit. Andromonoecious cultivars generally have staminate flowers in axillary clusters on the main stem and laterals, and a single perfect flower at the first node of each lateral branch. One or two fruit are usually produced per plant, but cultivars with small fruit may have more. Immature fruit are pubescent; mature fruit are glabrous. The small white to tan seeds are oblong to elliptical.

The fruit of different cultivar types are quite diverse, making melon the most variable species in its genus. Some cultivars have fruit with a reticulate netting on the rind (Fig. 4.4; see also Fig. 4.11) whereas others have smooth rind. Indented vein tracts distinguish various cultivars, but these sutures are lacking in other cultivars. Rind colour varies from green, sometimes with white stripes, to yellow, tan or white when mature (Figs 4.5, 4.6, 4.7). Flesh colour may be orange, pink, green or white (Figs 4.7, 4.8). An abscission layer forms at the attachment of the fruit to the peduncle in some cultivars, causing the fruit to separate from the vine at maturity. In other cultivars, the peduncle is persistent.

Melon plants will not cross with cucumber plants, although these species are members of the same genus. The mistaken belief that cucumber and melon

Fig. 4.4. 'Dulce' western shipper reticulatus melon with dense rind netting.

Fig. 4.5. Fruit of 'Golden Beauty' melon (*Cucumis melo*), showing yellow rind (top) and white flesh (bottom).

Fig. 4.6. F_2 population of *Cucumis melo* var. reticulatus × *Cucumis melo* var. agrestis, demonstrating genetic variation.

Fig. 4.7. Fruit of oriental crisp-flesh melon (*Cucumis melo*), showing the exterior radial cracking (top) and white interior (bottom).

Fig. 4.8. 'Chujuc' western shipper melon with high beta-carotene.

are able to cross may have arisen from harvesting melons when immature or after their quality has been impaired by disease or unfavourable climate, and assuming that the insipid flavour of the melon was due to cross-pollination with a cucumber. Cucumber flowers pollinated by melon pollen will begin to develop, but the embryo aborts before it can be rescued by embryo culture techniques. Seedless cucumbers may be produced from cross-pollination with melon pollen, but melon fruit generally do not develop when pollinated with cucumber pollen.

Origin

Recent molecular phylogenetic analysis of Australian and south Asian *Cucumis* species suggests that *C. melo* likely originated in India from *C. trigonus* or *C. callosus* (Sebastian *et al.*, 2010; Endl *et al.*, 2018). Sebastian *et al.* (2010) also found that the Australian species *C. picrocarpus* has the highest DNA sequence similarity to *C. melo*. However, the oldest wild relatives of melon in the subgenus *Melo*, including *C. hirsutus*, *C. humifructus* and *C. sagittatus*, originated in Africa.

Historical records and archaeological remains place melon in Egypt and Iran during the 2nd and 3rd millennia BC, respectively. Melon spread throughout the Middle East and Asia, becoming an important vegetable in India, Egypt, Iran and China. Iran, Afghanistan and China are considered secondary centres of melon diversification. Genetic diversity in Spain is great as well.

In the 15th century, melon was brought from Turkish Armenia to the papal estate of Cantaluppe, near Rome, and was distributed from there to western Europe. The word 'cantaloupe' is derived from this source. However, the small melon with a hard rind grown in Italy at that time is quite unlike the modern cultivars known as cantaloupes today. Ancient Romans who grew melon did not prize the fruit as much as they did cucumbers, presumably due to the poor quality of the melon cultivars they grew. In the 17th century, melon cultivars were under cultivation in British glasshouses.

Melon was brought to the New World by Columbus and carried to California by the Spaniards in 1683. Most of the early American cultivars had green flesh. 'Rocky Ford', a selection of the green-fleshed 'Netted Gem', became important in the late 19th century. Orange-fleshed cultivars such as 'Bender' and 'Irondequoit' were popular around 1900. Improved orange-fleshed cultivars, including 'Hearts of Gold' (introduced in 1914) and 'Hales Best' (1924), were predominant for many years until they were replaced by cultivars resistant to powdery mildew in the 1930s. Although the first F_1 hybrid melon, 'Burpee Hybrid', was introduced in 1955, introduction of hybrid cultivars of melon proceeded at a relatively slow pace until the 1980s. Hybrid cultivars, because of superior performance and disease resistance, are now dominant (Fig. 4.9).

Melon is now grown in temperate and tropical areas throughout the world. China produces more than twice as much as any other country (see Chapter 2, Table 2.1). Melons are so prized in Japan that they are grown in glasshouses and command prices many times those prevailing in other countries. In Japan, the greenhouse melons are often given as gifts. Western Asia (e.g. Turkey and Iran) and northern Africa (e.g. Egypt and Morocco) are also important areas for melon production. Spain, Romania, Italy and France are the largest European producers (FAO, 1995).

Fig. 4.9. 'Doublon' Charentais melon of the cantalupensis type, resistant to *Fusarium oxysporum* f.sp. *melonis*, races 0, 1.

Classification, cultivars and uses

French botanist Charles Naudin determined in the mid-19th century that several *Cucumis* species would cross readily with each other and with melon, and he combined them as different botanical varieties of *C. melo*. Today, with slight modification, Naudin's categories are considered horticultural groupings based on fruit characteristics and uses, and not botanical varieties based on phylogeny (Munger and Robinson, 1991), nor conforming necessarily with taxonomic rules for nomenclature (Trehane *et al.*, 1995). These cultivar groups are as follows.

1. Cantalupensis Group (cantaloupe, muskmelon). Medium-sized fruit with netted, warty or scaly surface; flesh usually orange but sometimes green; flavour aromatic or musky. Fruit detach from peduncle at maturity. Usually andromonoecious.
2. Inodorus Group (winter melon). Fruit usually larger, later in maturity and longer shelf life than those of the Cantalupensis Group. Rind surface smooth or wrinkled, but not netted; flesh typically white or green and lacking a musky scent (Fig. 4.10). Fruit do not detach from the peduncle when mature. Usually andromonoecious.
3. Flexuosus Group (snake melon). Fruit very long, slender and often ribbed. Used when immature as an alternative to cucumber. Monoecious.
4. Conomon Group (pickling melon). Small fruit with smooth, tender skin, white flesh, early maturity and usually with little sweetness or scent. Often pickled and also eaten fresh or cooked. Andromonoecious.
5. Dudaim Group (pomegranate melon, Queen Anne's pocket melon), including members of the previously recognized Chito Group (mango melon, vine peach). Small, round to oval fruit with white flesh and thin rind.
6. Momordica Group (phoot, snap melon). Fruit ovoid to cylindrical in shape, with dimensions of 30–60 × 7–15 cm. Flesh white or pale orange, low in sugar content, mealy, and insipid or rather sour tasting. Smooth surface of fruit cracks or bursts as maturity approaches and fruit disintegrates when barely ripe. Most cultivars monoecious.

Fig. 4.10. 'Honeydew' inodorus melons with distinct mature rind colours.

These cultivar groups belong to ssp. *melo*. Although plants of ssp. *agrestis* (Naud.) Pang. are mostly wild slender vines with small inedible fruit, they are sometimes cultivated in Asia and occur as weeds in many tropical areas. *C. melo* ssp. *agrestis* has been found to possess some valuable disease resistance traits, which are being introgressed into cultivated melon.

Cantalupensis Group cultivars are the most important in commerce. Cultivars grown in the western USA that are called cantaloupes produce moderately ribbed, tightly netted fruit that ship well and have thick, orange flesh. 'PMR 45' was introduced by the USDA and the University of California in 1936, and many selections of this cultivar were subsequently made by breeders. Because of resistance to powdery mildew, 'PMR 45' and its derivatives dominated California melon production for more than half a century. 'Persian' and related cultivars, which are also grown in California, are similar to 'PMR 45', but have larger fruit with heavy netting (Fig. 4.11). Cultivars grown in the eastern USA may also produce large fruit, but generally with less netting and more ribbing (e.g. 'Iroquois'). High-quality melons of the Charentais type (e.g. 'Charentais', 'Charmel', 'Alienor'), with thin rinds and only sparse netting, are popular in Europe and Africa. Green-fleshed netted melons developed in Israel (Galia types) are also popular in Europe.

Breeders have surveyed and selected for various traits in Cantalupensis Group cultivars. Leading harper-type cultivars in the USA such as 'Athena,' 'Caribbean Gold,' 'Sweet Spring' and 'Fiji', and eastern-shipper types 'Accolade' and 'Astound' are tolerant of environmental conditions detrimental to many other cultivars: they can be grown on saline soil; they tolerate high concentrations of atmospheric ozone; and they can be treated with sulfur at high temperature for disease control without injury. Melon cultivars have also been found to differ in sensitivity to high boron concentrations in soil. Birdsnest-type cultivars from Iran, which have a compact plant habit, reduced apical dominance, good fruit set and concentration of maturity, are potentially valuable for once-over harvesting (McCollum *et al.*, 1987). 'Siberian Honey Dew', an Inodorus Group cultivar, has been used as a parent to breed

Fig. 4.11. 'Uzbek' melon, reticulatus type with very large fruit size, 6 kg.

long storage life of fruit into cultivars of the Cantalupensis Group. This trend has accelerated dramatically during the past 10 years, with many commercial programmes utilizing Inodorus germplasm to cross with Cantalupensis lines, generating a greater diversity of low ethylene-producing or ethylene-insensitive netted melons. Many of these novel orange-fleshed cultivars, such as 'Caribbean Gold', 'Fiji', 'Karameza', 'Infinite Gold' and 'Sweet Spring', have taken a substantial share of the cantaloupe market away from the traditional western and eastern shipper cultivars. Extremely firm flesh and low ethylene production may allow for longer field and shelf life, reduced harvest expense, and more consistent quality to the consumer. However, aromatic volatiles associated with flavour have been diminished, despite high sugar content (high °Brix) and high beta-carotene level. Texture may also be excessively crisp or rubbery, depending on the genetics and the production environment.

Cultivars of the Inodorus Group are called winter melons because the fruit generally store well. Some are shipped to distant markets from Spain, California and other areas with a long warm growing season. 'Honey Dew', a classic member of this cultivar group, was introduced to the USA from France at the beginning of the 20th century. The firm fruit has green, sweet flesh. It is generally grown in the western states of the USA and other long-season areas, because of its late maturity. Bush habit has been backcrossed into 'Honey Dew' and related cultivars. 'Piel de Sapo' from Spain is another important melon cultivar of this group. The winter melon cultivar 'Casaba' has a yellow, furrowed rind and white flesh of high quality. It also requires a long growing season. 'Juan Canary' fruit are yellow and elongated, with light green flesh. 'Crenshaw' has delicious juicy orange flesh, but does not ship or store as well as 'Honey Dew', 'Casaba' or 'Juan Canary'. 'Honey Gold' is one of the more recently created orange-fleshed winter melon cultivars. Various Spanish landraces (e.g. 'Común') of the Inodorus Group have resistance to powdery mildew (Floris and Alvarez, 1995).

Some melons of the Flexuosus Group have fruit that are long, for example 150 cm in 'Cooking Queen'. The fruit are slender and coiled (less so when plants are trellised), hence their common name of snake melon (Fig. 4.12). The 'Armenian Cucumber' cultivar listed in seed catalogues is not a cucumber, but a melon of this cultivar group. Snake melon is popular in the Middle East, where it is harvested when immature and consumed like cucumbers. It is better than cucumber plants at setting fruit at the high temperatures often prevailing in these countries.

Pickling melon cultivars of the Conomon Group produce small, crisp fruit with white flesh and often a low sugar content (Fig. 4.13). This lineage is believed to have originated in China and Japan more than 1000 years ago. The fruit are pickled there and some think that the melons make better quality pickles than cucumbers. They have a thin tender rind, which can be white (e.g. 'Shiro-Uri') or green with stripes. Cultivars such as 'Golden Crispy' may be eaten fresh with skin intact, like an apple. Young fruit are also cooked, in

Fig. 4.12. Fruit of a Flexuosus Group melon measuring approximately 1 m long.

Fig. 4.13. Pickling melons (*Cucumis melo*) used for making pickles primarily in Asia with long (left), round (centre) and short (right).

ways similar to summer squash. Resistance to CMV has been found in these cultivars and breeders are attempting to transfer resistance to members of the Cantalupensis Group.

Some Dudaim Group cultivars are grown as ornamentals because of their aromatic fruit, but others lack this fragrance. The bland fruit are eaten fresh or pickled. The flesh resembles that of cucumber in colour and texture. Plants of this Asian cultivar group have escaped from cultivation and become established as weeds in southern USA and elsewhere.

Momordica Group cultivars are grown in India and other Asian countries, where the fruit are cooked as a vegetable, consumed fresh, or used in salads. The seeds are roasted and eaten in India. The USDA plant introduction accession PI 175111 has been useful in melon breeding programmes because of its resistance to aphids and viruses. It was originally collected from India from a street vendor selling seeds for roasting, and included both cucumber and melon seeds.

Melon seeds that have been dried and ground are commonly eaten in Africa, where they can be found in commercial trade among countries. Oil is pressed from the seeds and used in lamps for illumination. The leaves are sometimes prepared as a vegetable and the whole plant provides fodder.

In addition to food, various parts of the melon plant have been used medicinally. In China, fruit and roots are taken as a diuretic, roots and flowers as an emetic, leaves and seeds for the treatment of haematoma, and stems for the treatment of dysentery and hypertension. Melon plants contain various bioactive principles, including elaterin, stigmasterol, spinosterol and the antitumour cucurbitacin B (Duke and Ayensu, 1985).

SQUASH AND PUMPKIN

Botany

In *Cucurbita*, there are five domesticated (*C. argyrosperma, C. ficifolia, C. maxima, C. moschata, C. pepo*) (Table 4.1) and ten wild species (Schaefer and Renner, 2011). *C. ecuadorensis* may be an incipient domesticate, having begun the domestication process but later abandoned when *C. maxima* was introduced. Some accessions of *C. ecuadorensis* possess relatively large and non-bitter fruit, but the species is no longer actively cultivated in its centre of origin. Most species, including the domesticated species, are mesophytes with fibrous root systems; the remaining taxa are xerophytic perennials with enlarged roots. All produce frost-sensitive, tendril-bearing vines. Wild species and most cultivars have long trailing vines. However, some cultivars, particularly in *C. pepo*, have a compact bush habit in which the tendrils have been reduced in size and function. The large leaves are palmately lobed to nearly round. Some cultivars of *C. pepo* and *C. maxima*, as well as many cultivars of *C. moschata* and *C. argyrosperma*, have mottled leaves, with white or silvery areas at the junctions of principal veins.

Species of *Cucurbita* are monoecious. The unisexual flowers are large, showy and orange (except for the cream-coloured flowers of *C. okeechobeensis*), usually occurring singly in leaf axils. The five petals are reflexed at their tips and fused at their bases. Calyx lobes are narrow (*C. pepo*) or broad to sometimes leaf-like (*C. moschata*). Petals alternate with sepals, which are also united at their bases and fused with the lower corolla to form a cup-like hypanthium. Although the three filaments are separate, the anthers are more or less united and produce abundant amounts of heavy, sticky pollen. Styles are typically joined together, but they diverge slightly where the stigmas are attached. Nectar is produced in a disc inside and at the base of the hypanthium

Table 4.1. Distinguishing characters for domesticated species of squash (*Cucurbita*).

Species	Peduncle	Stem	Leaf	Seed
C. argyro-sperma	Hard, angular but becoming round at maturity, corky in the valleys, only slightly flared at fruit attachment	Hard, angular, grooved	Moderately lobed, softly pubescent	Usually white, may be very large; surface smooth or split; margin prominent, smooth to ragged, and sometimes dark
C. ficifolia	Hard, smoothly angled, with slight flaring	Hard, smoothly grooved	Lobed, nearly round, slightly prickly	Usually black, sometimes tan; surface often minutely pitted; margin smooth and narrow
C. maxima	Generally soft, round, often corky, not flared	Soft, round	Usually unlobed, nearly round, soft	White to brown, often plump; surface sometimes split or wrinkled; margin very narrow; seed scar oblique
C. moschata	Hard, smoothly angled, pentagonal in cross-section, broadly flared	Hard, smoothly grooved	Nearly round to moderately lobed, soft	Dull white to brown; surface smooth to somewhat rough; margin prominent, typically ragged, and often dark; seed scar slightly oblique
C. pepo	Hard, angular, sometimes slightly flared right next to fruit	Hard, angular, grooved, prickly	Palmately lobed, often deeply cut, prickly	Dull white to tan; surface smooth; margin prominent but usually smooth; seed scar square or rounded

(see Chapter 1, Fig. 1.2). The unilocular inferior ovary has three or five placentae, corresponding to the number of bilobed stigmas. The fruit is a pepo and there is a great diversity in fruit size, shape and colour among cultivars (Fig. 4.14). Fruit flesh is bitter in wild taxa and in most ornamental gourds of *C. pepo*. Seeds of the domesticated species are large, measuring up to 3 cm long in *C. argyrosperma* (Fig. 4.15).

Fig. 4.14. Four domesticated species of squash (left to right): *Cucurbita moschata* 'cheese' and 'butternut' types; *C. maxima* (behind) 'hubbard' type; *C. argyrosperma* 'cushaw' type; and *C. pepo* 'delicata', 'acorn' and 'jack o'lantern' (behind) types. Note differences in peduncle shape and size.

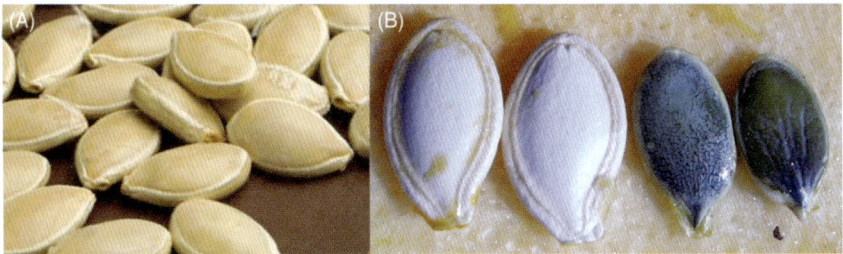

Fig. 4.15. Pumpkin seed used for culinary purposes. (A) *Cucurbita maxima* 'Golden Delicious' seed. (B) *C. pepo* hulled (left) and hull-less (right) pumpkin seed.

Fruit of the wild species have hard, lignified rinds which help to protect the seeds from herbivores. The fruit may remain intact long after the plant has died. After lengthy storage, little is left except for the dried rind, peduncle and seeds; these fruit parts may persist for centuries, enabling archaeologists and botanists to determine prehistoric species distributions and uses. The intact dried fruit are buoyant, permitting seed dispersal via waterways.

Nomenclature

Different *Cucurbita* species are known by the same common name; and different common names, such as squash, pumpkin, cushaw and gourd, have been used for the same species. This chapter will use the unqualified term squash to refer to any or all of the domesticated species of *Cucurbita*.

Varieties commonly called pumpkins are generally coarse-grained and strongly flavoured and may be used for pies, stock food, or ornament. Winter squashes are usually fine-grained, milder in flavour and used for cooking. Summer squashes are those forms, almost always *C. pepo*, used immature as a table vegetable.

The word squash is derived from the American aboriginal word 'askutasquash', meaning eaten raw or uncooked. Cultivars are classified as summer squash (sometimes called vegetable marrow) or winter squash, depending on whether the fruit is used when immature or mature. The term winter squash refers to the ability of the fruit to be stored until the winter months. Although summer squashes are generally *C. pepo*, a few cultivars of *C. moschata* and *C. maxima* have been developed for use as summer squash. Winter squashes may be *C. pepo* (e.g. acorn, delicata, jack-o'-lantern, Styrian oilseed and spaghetti types) (Fig. 4.16a), *C. maxima* (hubbard, kabocha and buttercup types), *C. moschata* (butternut type) or *C. argyrosperma* (e.g. 'Green Striped Cushaw)'. In *C. pepo*, bush habit distinguishes most summer squash cultivars (Fig. 4.16b) from winter squashes although increasing numbers of small winter squash and ornamental pumpkins have bush or semi-bush habit.

Pumpkin comes from the old English word 'pompion', the Greek 'pepon' and the Latin 'pepo', which together mean a large, ripe, round melon or gourd. Today, the term pumpkin is used in various ways and has no botanical meaning. It typically refers to any squash used for pies, jack-o'-lanterns or stock feed. *C. maxima* and *C. moschata* cultivars that would be called winter squashes in the USA are often called pumpkins in India and other countries.

Cushaw defines a winter squash cultivar with a curved neck. Its use is not limited to a single species. Thus, 'Green Striped Cushaw' is *C. argyrosperma*, but 'Golden Cushaw' is *C. moschata*.

Gourd often designates a cucurbit not used for food, e.g. wild species of *Cucurbita*. The cultivated ornamental gourds of *C. pepo*, which have small fruit of a wide assortment of shapes and colours, are used for decoration (Fig. 4.17). *Cucurbita* gourds have hard shells and some consider this a distinction between gourd and squash. However, the fruit of some summer squash cultivars also have a hard rind when mature.

Other terms for squash and gourds are calabaza and calabash, both deriving from the same Persian root. In general, calabash refers to hard-shelled gourds of several species (some of which are not in the genus *Cucurbita* or even in the Cucurbitaceae). Calabaza was applied soon after the Columbian exchange to New World cucurbits, but is most commonly applied to large *C. moschata* tropical pumpkins nowadays.

Origin and history

On the basis of isozyme banding patterns, Puchalski and Robinson (1990) classified *Cucurbita* species into seven phylogenetic groups. *C. moschata* and

Fig. 4.16. (a) An assortment of *Cucurbita pepo* winter squash types. (A) Jack-o'-lantern ornamental and pie pumpkins. (B) Spaghetti squash. (C) Acorn squash. (D) Delicata. (E) Styrian oilseed pumpkin. (F) Delicata–acorn type. (b) Summer squash (*C. pepo*) types: (left to right) zucchini, bicolour straightneck, crookneck, ball (top) and patty pan (bottom).

Fig. 4.17. Diversity of *C. pepo* gourds.

C. argyrosperma were included in the same group and the other cultivated species were placed into different groups. In spite of their occurrence in the same phylogenetic lineage, *C. moschata* and *C. argyrosperma* were probably domesticated from distinct wild progenitors, as were the other cultivated squashes. The original phylogeny proposed by Puchalski and Robinson (1990) is supported by genome sequence analysis (Montero-Pau *et al.*, 2017).

Most *Cucurbita* species originated in North or Central America, but a few species, including *C. maxima*, are native to South America. Since the wild fruit of this genus are bitter and inedible, early gatherers probably first collected the fruit for their edible seeds or to use the durable rinds as containers. These uses eventually gave way to the domestication of species with edible fruit. From these beginnings, squash, along with maize and beans, became a staple in the diets of the Aztec, Incan and Mayan civilizations of Latin America.

Archaeological evidence places wild populations of *C. pepo* in Mexico and eastern USA around 5000–10,000 years ago (Smith, 1997a). Domestication of this species apparently took place independently in these two areas (Decker, 1988) with domestication in eastern USA occurring about 5000 years later than the Mexican domestication (Smith, 2006). The US cultivars of ssp. *ovifera* (L.) Decker var. *ovifera*, which include various summer squashes and most ornamental gourds, were probably selected mainly from wild populations of var. *ozarkana* Decker-Walters inhabiting the Mississippi Valley. However, populations in Texas (var. *texana* (Scheele) Decker) and northeastern Mexico (ssp. *fraterna* (Bailey) Andres) may have contributed to the genetic evolution of these cultivars as well. The wild progenitor of the Mexican lineage of cultivars (ssp. *pepo*), which includes pumpkins and vegetable marrows, is currently unknown and possibly extinct (Decker-Walters *et al.*, 1993). Interestingly, there was apparently no introgression of genes between *C. pepo* var. *pepo* and *C. pepo* var. *ovifera* over the course of a few thousand years, even though there must have been human migrations and the regions of domestications were not that far apart.

C. pepo was the first squash introduced to Europe. It appears in the Livre d'Heures d'Anne de Bretagne illustrated between 1505 and 1508 (Paris et al., 2006). Images of fruit and flowers of gourd forms of C. pepo are present in the Villa Farnesina in Rome in festoons of vegetation separating spandrels, and were illustrated between 1509 and 1517 (Janick and Paris, 2006). It was not until Fuchs' Di Historias Stirpium was published (1543) that any of the Cucurbita species appeared in herbals. Some of the fruit portrayed in ancient European herbals are not unlike those of modern cultivars. Exquisitely detailed images that provide insight into the diversification of C. pepo in Europe and the origins of summer squash forms can be seen in the paintings of Antoine Nicolas Duchesne (Paris, 2000). Secondary diversification of pumpkin and vegetable marrow cultivars occurred in Asia Minor (Paris, 2001). Today, C. pepo is grown throughout the world.

C. moschata was cultivated in Mexico, South America and southwestern USA in pre-Columbian times. It may have been domesticated independently in Mexico and northern South America. However, the wild ancestor of C. moschata is currently unknown. As a cultivated plant, this species migrated throughout the Caribbean islands, giving rise to various indigenous 'calabaza' landraces. When it reached Florida, Native Americans developed a distinct landrace called 'Seminole Pumpkin'. Additional diversification of cultivars has taken place in several areas of Asia and Africa. Images of C. moschata in the Villa Farnesina dating to the early 16th century have been reported (Janick and Paris, 2006), but in general, there are fewer illustrations of this species in the artwork and herbals in Europe.

C. argyrosperma was apparently domesticated in southern Mexico, where archaeological evidence of cultivation dates back 5100 years (Smith, 1997b). Wild populations of ssp. sororia (Bailey) Merrick & Bates, which occur in Mexico and Central America, gave rise to the domesticated subspecies (ssp. argyrosperma). Landraces of ssp. argyrosperma var. argyrosperma may have been selected first. The other two groups of cultivars, var. callicarpa Merrick & Bates and var. stenosperma (Pang.) Merrick & Bates, probably evolved from northern and southern landraces of var. argyrosperma, respectively (Merrick, 1990). Weedy populations of var. palmeri (Bailey) Merrick & Bates are considered escapes of var. callicarpa that may have undergone introgression with ssp. sororia. Today, C. argyrosperma is cultivated in Mexico, Latin America, the Caribbean, southwestern USA and Asia, but is less commercially important than C. pepo, C. maxima or C. moschata.

C. maxima ssp. andreana (Naud.) Filov, a bitter-fruited native of Argentina and Uruguay, is the wild progenitor of domesticated ssp. maxima and the two taxa hybridize readily. This species was first encountered by the Spanish when their explorations reached the Pacific coast of South America in the 16th century, and the species is apparently illustrated in the Villa Farnesina (Janick and Paris, 2006; Formiga and Myers, 2019). *C. maxima* types did not reach North America until they were brought by sailing ships from South America to northeastern USA in the 18th century. One of the early introductions was called 'Valparaiso', presumably having originated from that city in Chile. Well adapted to temperate climates, *C. maxima* types have been especially popular in the northern tier of states in the USA, Canada and Europe. Today, *C. maxima* is cultivated worldwide, with significant production in South America, India and Africa. Kabocha (buttercup) squash is a huge export item to Asia for both Mexico and New Zealand, and *C. maxima* 'Jarradale' (crown-type blue/grey squash) is the major winter squash grown for local consumption in New Zealand and Australia. There is also widespread cultivation in Turkey; and in Poland, *C. maxima* breeding has been ongoing for some years, including the use of varieties as a forage for cattle. Because this species is generally more cold tolerant than *C. argyrosperma*, *C. moschata* and *C. pepo*, it can be grown at higher latitudes and elevations.

    C. ficifolia is called the Malabar gourd because it once was thought to be of Asiatic origin. However, like other squashes, its origin is in the Americas, probably in the high-altitude mountain areas of Mexico. This cold-tolerant species, growing at elevations of up to 2600 m, was also cultivated in the Andes Mountains of Peru in prehistoric times. It reached Asia on sailing ships shortly after the Columbian exchange began. The fruit have hard rinds and long shelf life (storage for up to 2 years) and could withstand the long voyages. Malabar gourd is seldom grown in industrialized countries, except as a rootstock on which to graft cucumber. However, in Latin America, the species provides a vegetable called 'zambo' as well as food for livestock. Even the fruit of wild plants are gathered and eaten (Andres, 1990).

Species descriptions

Squash species can be distinguished on the basis of peduncle, foliage and seed characteristics (see Table 4.1). The peduncle is particularly distinctive for each of the domesticates (Fig 4.18).

    C. pepo is a highly polymorphic species. Types include various summer squashes (vegetable marrows, crooknecks, straightnecks, zucchinis, cocozelles and scallops), ornamental gourds, winter squashes (acorns and delicatas), pumpkins (pie, jack o' lantern, and Styrian oilseed) and unique cultivars such as 'Vegetable Spaghetti'. Its fruit types are more genetically diverse than those of any other species in the Cucurbitaceae. They are green, yellow, orange, white, striped or variegated; smooth, ribbed, furrowed or warty; and flat, round, oval, elongated, necked or otherwise shaped. Fruit size ranges from small (5 cm in diameter in some ornamental gourds) to large (over 50 cm in diameter in some pumpkins). Flesh colour is usually white for summer squash and orange for winter squash, but often of a lighter intensity than the orange flesh colour of

Fig. 4.18. Peduncles of four domesticated species of *Cucurbita*. (A) *C. pepo* (Mississippi river valley centre of domestication). (B) *C. pepo* (Mexican centre of domestication). (C) *C. maxima*. (D) *C. argyrosperma*. (E) *C. moschata*. Note that *C. pepo* peduncles are hard, star-shaped in cross-section and not flared at the base. *C. maxima* peduncles are large, smooth and spongy. *C. argyrosperma* peduncles appear similar to *C. maxima*, but have hard ridges with spongy tissue filling in the valleys. *C. moschata* peduncles are hard, pentagonal in cross-section shape and flared at the base.

winter squashes of *C. maxima* and *C. moschata*. The peduncle on *C. pepo* cultivars is generally hard and angular with five acute valleys and ridges. Within this species the two centres of domestication can be distinguished by appearance of the peduncle, among other traits. Relatively smaller, long and gracile peduncles are found on cultivars originating from the Mississippi River Valley centre of domestication whereas larger, more robust peduncles are indicative of the Mexican centre (Fig. 4.19).

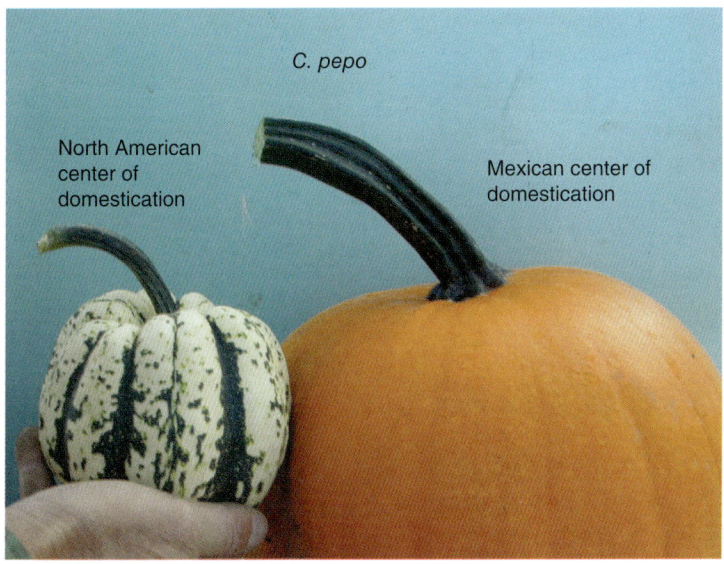

Fig. 4.19. *Cucurbita pepo* from North American (left) and Mexican (right) centres of domestication showing differences in peduncle size.

Cultivars of *C. pepo* ssp. *ovifera* generally have smaller reproductive and vegetative parts than those of ssp. *pepo*. Fruit of wild populations are small, bitter, round to pyriform, and typically white (ssp. *ovifera* var. *ozarkana*), green-and-white striped (var. *texana*), or green-and-white striped turning yellow-orange when mature (ssp. *fraterna*).

Some *C. maxima* cultivars produce the largest of all squashes. In giant pumpkin contests, the winning fruit in 2016 weighed 1193 kg. Fruit of this species are: (i) orange, green or grey; (ii) smooth or ribbed; (iii) smooth or warted; (iv) typically round or oval; and (v) sometimes with a protuberance at the blossom end. This reaches an extreme in turban types where about half of the ovary is exposed. Seeds are usually quite large and plump, white or brown, and rugose or smooth. Peduncles on *C. maxima* types tend to be large, round in cross-section and spongy, initially soft but hardening on fruit held in storage. While most cultivars are used for winter squash, a few are eaten as an immature vegetable and the seeds may be consumed. 'Golden Delicious' is a large hubbard type with soft orange rind that is used both for processing of its flesh into a purée as well as for its large plump white edible seeds (Fig. 4.20)

C. moschata cultivars produce fruit that are often large (up to 25 kg); smooth or ribbed, with smooth, warted, or wrinkled rinds; typically mottled but often green or buff-coloured although naturally occurring yellow-skinned types are known (the precocious yellow gene *B* from *C. pepo* and B^{max} from *C. maxima* have also been introgressed). Landraces have flattened, round, oval,

Fig. 4.20. 'Golden Delicious' hubbard type *C. maxima* pumpkin used for processing and culinary seeds.

or curved-neck fruit. Seed colour can range from very dark brown (nearly black) to almost white. Flesh colour ranges from yellow to orange, and some Central American types have a dark green pigment in the placenta and adjacent pericarp. Breeders have developed types having a long, straight neck with the small seed cavity confined to the enlarged blossom end of the fruit, e.g. 'Waltham Butternut' and 'Ultra'. Winter squash cultivars of this species store well and can be of very high quality.

C. *argyrosperma* is a squash species used for food in Mexico, Latin America and southwestern USA for centuries but was not recognized as a distinct species until 1930, when a Russian botanist named it *C. mixta* Pangalo. Before then, it was included in *C. moschata*, a species morphologically similar, but differing in biochemical markers and separated by sterility barriers. *C. mixta* was reclassified as *C. argyrosperma*, and the former species *C. palmeri* Bailey and *C. sororia* Bailey were designated as infraspecific taxa of *C. argyrosperma* (Merrick and Bates, 1989). Variety *argyrosperma* and var. *stenosperma* have round to pyriform cream-coloured fruit with green-mottled stripes. Fruit of var. *callicarpa* are more variable in shape and coloration. Those of var. *palmeri* and ssp. *sororia* are relatively small, often with bitter flesh, especially in ssp. *sororia*. The pale yellow to orange flesh of the cultivars is coarse and often of rather poor quality for a winter squash. A dark green tint to the placental tissue is associated with ssp. *argyrosperma* var. *stenosperma* (the trait is also found in *C. moschata*). Variety *argyrosperma* often has very large seeds characterized by wing-like margins, perhaps allowing easier decortication of seeds that are often eaten as 'pepitas' in Mexico and Guatemala. Variety *stenosperma* is also grown primarily for its edible seeds. Seeds of var. *callicarpa* are variable in colour and surface characteristics. The tan seeds of ssp. *sororia* are small, whereas the white seeds of ssp. *argyrosperma* var. *palmeri* are somewhat larger. Peduncles of immature fruits of *C. argyrosperma* resemble those of *C. moschata* while peduncles of mature fruits of *C. argyrosperma* superficially resemble those of *C. maxima*. Mature fruit peduncles of *C. argyrosperma* tend to be somewhat smaller with the valleys between the hard ridges of parenchyma tissue filled in with softer, spongy callus.

The species name for *C. ficifolia* and its common name of fig leaf gourd refer to the resemblance of its leaves to the shape of fig tree leaves. Despite its long history of cultivation, there is little diversity of fruit type in Malabar gourd. The fruit can be solid white, or green-and-white mottled, with white, coarsely fibrous flesh (Andres, 1990). In shape, the fruit resembles a watermelon. No other wild plants of a domesticated squash species produce fruit as large as those of feral Malabar gourd. This is the only species of *Cucurbita* with black seeds, although landraces with tan seeds also occur. Malabar gourd is a short-day plant and not particularly early flowering outside the tropics. However, USDA National Plant Germplasm System (NPGS) germplasm accessions have been grown successfully to maturity in Western Oregon at 45° N latitude. Breeders producing seed of this species for use as rootstocks have selected for earlier flowering day-neutral forms, but even these selections are late to mature.

Uses

During domestication in pre-Columbian times, people selected for non-lignified fruit rinds in winter squashes and pumpkins, which were usually consumed when ripe. Hard rind, which is conditioned by a single dominant allele in *C. pepo*, is found in many cultivars of summer squash that are consumed before rind lignification becomes objectionable.

Native Americans dried strips of squash flesh in the sun for preservation. Today, summer squash is usually cooked by boiling or frying and winter squash by baking, boiling or microwaving. Summer squash generally has white flesh and low dry matter content, but winter squash has been bred to have orange carotenoid-rich flesh high in dry matter.

Immature fruit of winter squash cultivars like members of the acorn group (*C. pepo*) may be eaten like summer squash. 'Butternut' and similar cultivars of *C. moschata* are of high quality when immature. However, they are seldom marketed this way, due to relatively low yield, late maturity and the vine habit, which makes harvesting difficult. In Southern Europe, *C. moschata* was selected as a summer squash resulting in 'Tromboncino' landraces. These types retain a vine habit but are prolific in production of slender long-necked fruit. The immature fruit is normally eaten when 15–20 cm in length. At maturity, the fruit may be up to 1 m in length. The eating quality of 'Tromboncino' types at the immature stage is considered by many to be superior to *C. pepo* summer squash. They are higher in total solids, have flavour more akin to a mature winter squash and the fruit possess a long seedless neck. In Argentina, bush cultivars of *C. maxima* known as 'zapallito' are a popular form of summer squash.

Some summer squash cultivars, e.g. the vegetable marrows (*C. pepo*), are consumed when almost mature. In the Middle East, nearly mature fruit of 'Cousa' are stuffed with meat and other ingredients, then baked.

In the USA, the most familiar use of pumpkins of *C. pepo* is for Halloween jack-o'-lanterns. Every autumn, the orange, round or oval fruit are carved into grotesque faces and illuminated from within by candles.

Commercially canned pumpkin pie mix may be made of *C. maxima* or *C. moschata*. Historically, *C. pepo* has had a long tradition of use in the USA, but fruit of the other two species produce better baked pies and *C. pepo* is no longer used. Cultivars of *C. moschata* such as 'Ultra' butternut, 'Dickinson Field' and *C. maxima* such as 'Golden Delicious' are also processed and sold as canned or frozen winter squash. Canners prefer cultivars with an orange or tan rind. If a small piece of the skin is inadvertently included in the canned product, it is less noticeable than with green-fruited cultivars. Types used for processing have soft rinds with total solids from 5% to 9% and soluble solids from 6% to 12%. Orange carotenoid-rich fruit of *C. maxima* that are also fine in texture and flavour are mashed into canned baby food.

Although banned in the USA since the 1970s, the organophosphate pesticide DDT and its breakdown products dichlorodiphenyldichloroethylene (DDE) and dichlorodiphenyldichloroethane (DDD) will persist in soils for decades. Other chlorinated hydrocarbon pesticides may include lindane, aldrin and dieldrin. *Cucurbita* species have extensive root systems and are known to be hyper-accumulators of DDE and DDD (White, 2002). Processors growing cucurbit crops on soils that have a history of DDT application are required to test for these compounds; if present in levels above FDA threshold of 0.1 ppm, they cannot sell processed product made from pumpkins grown in these fields. This is especially important for purée destined for baby food.

Mature fruit of *C. ficifolia* are used as winter squash and the immature fruit like summer squash. An Aztec candy called 'angel's hair' is prepared from the boiled stringy flesh. The bland fruit contain a proteolytic enzyme that may be of future commercial value to the food industry. Stem tips and leaves are used as greens, flowers are served as a condiment and seeds are roasted and eaten.

Squash was most likely domesticated originally for its protein- and lipid-rich seeds. In addition to being very bitter, wild cucurbits have thick inedible rinds and little edible pericarp, but seeds would have been readily accessible and relatively easy to process to remove their bitterness (Hart, 2004). The main effort required in processing contemporary squash seed is in removing the hull from the embryo, since bitterness has been bred out of most cultivars. In Latin America, hulled *C. argyrosperma* seeds are eaten as 'pepitas'. Hull-less *C. pepo* pumpkin seed derived from Styrian oilseed pumpkins are gaining popularity, because minimal preparation is required. *C. maxima* cultivars such as 'Golden Delicious' are an important snack food throughout Asia, where they are consumed in a manner similar to sunflower seeds in the USA.

Squash has been an important food in Mexico for centuries. Fruit are processed and consumed in a variety of ways. Squash seeds are a popular snack food and are also ground into a meal used to make special sauces. The flowers are eaten stuffed or fried, and lend colour and flavour to soups, stews and salads. Stems, growing tips and roots of some species are also consumed.

New World aborigines have used squash seeds as a source of cooking oil and medicinally as a diuretic, antipyretic and anthelmintic. The fruit flesh of wild species generates a saponin-rich froth when rubbed and has been employed to wash clothes.

Cultivars and breeding

Because of the great diversity of cultivars in *C. pepo*, various horticultural classifications have been proposed over the years. However, a nomenclaturally correct classification of cultivar groups (cf. Trehane *et al.*, 1995) has not yet been established for any squash species. For *C. pepo*, we used the classification by Paris (2008), which is based primarily on fruit shape:

1. Pumpkin. Fruit orange, round or oval. Small to large fruit, smooth to slightly ribbed.
2. Scallop. Fruit small, flattened, typically with scalloped edges.
3. Acorn. Fruit small, top-shaped, strongly ribbed and deeply furrowed, pointed at the blossom end.
4. Crookneck. Fruit elongated with a curved neck.
5. Straightneck. Fruit cylindrical with a straight, slightly constricted neck.
6. Vegetable marrow. Fruit short, cylindrical, tapering from the broad blossom end to the narrow peduncle end.
7. Cocozelle. Fruit long, cylindrical, tapering away from the large blossom end, with a length-to-width ratio of 3.5 or more.
8. Zucchini. Fruit long, cylindrical, with little or no taper.
9. Delicata. Fruit short, cylindrical with slight ribbing and blunt ends. Green striping on yellow or tan background.

The pumpkin, vegetable marrow, cocozelle and zucchini groups belong to ssp. *pepo* and the remaining groups represent ssp. *ovifera*. Not included above are the ornamental gourds, which are small when mature and occur in a variety of shapes and colours.

'Cocozelle', long grown in southern Europe, was introduced into the USA by the Asgrow Seed Company in 1934, and 'Caserta' by the Connecticut Agricultural Experiment Station in 1949. They are popular summer squashes with striped fruit. 'Caserta' produces a high proportion of female flowers early in the season, which is one reason for its use as the maternal parent of F_1 hybrid cultivars.

Dark green zucchini-type cultivars (e.g. 'Black Zucchini', introduced in 1931) are the most popular summer squashes in the USA, but lighter green cultivars (e.g. 'Cousa') are preferred in the Middle East. Zucchini cultivars have long been important in Italy and the earliest selections probably originated there by selection out of cocozelle types (Lust and Paris, 2016). F_1 hybrids are the predominant type grown today. Traditionally most zucchinis have green rinds, but researchers have bred types with the *B* allele for yellow rind ('Burpee's Golden Zucchini' being one of the first).

'Yellow Crookneck' and similar cultivars with a curved neck and hard, warty rind when mature are popular in southern USA. Cultivars of the straightneck group were probably originally selected from these crookneck squashes. 'Early Prolific Straightneck', an important yellow-coloured summer squash, was introduced by the Ferry Morse Seed Company in 1938. 'Yankee Hybrid', a straightneck squash introduced in 1942, was the first F_1 hybrid squash cultivar. 'Zephyr' is a bicolour straightneck × zucchini F_1 hybrid.

The fruit of 'White Bush Scallop' summer squash is similar to that depicted in ancient European herbals. It differs from the medieval squash, however, by having a bush habit. In addition to white, newer scallop squash cultivars may have bush habit, powdery mildew resistance and dark green or yellow rinds.

Trends in breeding zucchini and other summer squashes are to develop new hybrid cultivars with improved uniformity, better disease resistance and prolificacy, on an upright bush habit with open canopy and spineless petioles and leaves.

'Table Queen', an acorn type and one of the few winter squash cultivars of *C. pepo*, was introduced in 1913 by the Iowa Seed Company. This acorn squash produces small, ribbed fruit that are often cut in half and baked. It is similar to a landrace grown by Native Americans before the Columbian Exchange. 'Table Queen' and its antecedents have a vining habit, but 'Table Ace' and other bush cultivars with fruit shape similar to 'Table Queen' have been bred. Acorn squash is now one of the three primary squash types grown for large wholesale markets; modern cultivars, mostly hybrids, are bush or semi-bush (bush × vine crosses) in stature and have been bred with intermediate resistance to powdery mildew. Acorn squash can have some of the best culinary quality of the *C. pepo* winter squashes, but what is sold in many retail markets is picked before being fully mature and may be lacking in quality.

'Delicata', another winter squash, was introduced by the Peter Henderson Company in 1894 and is still grown today. The cylindrical fruit, which are cream-coloured with green stripes, are arguably the best culinary quality of the *C. pepo* winter squashes. 'Honey Boat' and 'Sugar Loaf' are tan-skinned Delicata releases from the Oregon State University programme and 'Zeppelin' is a re-selection of the landrace Delicata. Cornell University has released 'Bush Delicata', which has bush habit and powdery mildew resistance.

'Connecticut Field' pumpkin has been grown in the USA since colonial times and is still listed in some seed catalogues today. Miniature pumpkins of

C. pepo (e.g. 'Jack Be Little', 'Munchkin' and 'Baby Bear'), with fruit looking like jack-o'-lantern but much smaller, have become popular in recent years. They are used for decoration and are edible as well. 'Connecticut Field' is the forerunner to the rather large class of large ornamental jack-o'-lantern pumpkins used primarily for Hallowe'en. Many of these are F_1 hybrids with either bush or vining habit, and some have intermediate powdery mildew resistance. Traits of importance in this class include large slightly ribbed and dark orange fruit with large, sturdy and dark green peduncles for ease of handling.

Vegetable marrows, which are grown in the UK and elsewhere, are used at all stages of maturity. 'Vegetable Spaghetti', an unusual member of this group, is said to have originated in Manchuria and was introduced into North American commerce in 1936. When mature, the yellow, oval fruit can be cooked intact, after which, it is cut open to serve the thin strands of flesh resembling spaghetti. Scientists from Israel selected an orange hybrid cultivar called 'Orangetti' in 1986. In addition to having a higher concentration of carotenoids in the orange flesh, this cultivar differs from the vine-habit 'Vegetable Spaghetti' by having a semi-bush habit.

Oilseed pumpkins were cultivated over 200 years ago in Central and Eastern Europe. Originally hard-seeded, hull-less or naked-seeded types were identified in about 1880 in the Styrian region of Austria. Hull-less types do not require removal of the seed coat before crushing for oil extraction. A single allele plus modifier genes in *C. pepo* inhibit formation of the seed coat. Pumpkin cultivars such as 'Lady Godiva' and 'Triple Treat' have been bred in the USA to have hull-less seeds that are tasty, tender and nutritious, being high in protein and oil content. Newer hull-less seeded cultivars have semi-bush growth habit, higher seed yields and intermediate resistance to powdery mildew. Since the seed coat is maternal tissue, cross-pollination of a naked seeded cultivar will not affect that trait until the next generation.

Three horticultural groupings of *C. moschata* cultivars are recognized in the commercial trade of North America:

1. Cheese. Fruit variable, but usually oblate with a buff-coloured rind.
2. Crookneck. Fruit round at the blossom end with a long straight or curved neck.
3. Bell. Fruit bell-shaped to almost cylindrical.

These groups do not encompass all of the fruit types that have evolved in tropical America (e.g. calabaza landraces) and Asia. For example, Colombian landraces have small fruit with dark seeds, and Japanese cultivars (e.g. 'Chirimen', 'Kikuza') often have warty and wrinkled fruit.

The bell squash 'Butternut' is an important winter squash with excellent quality. It was selected for better fruit shape from the heirloom cultivar 'Canada Crookneck' and introduced by the Breck Seed Company in 1936. The elongated neck of the buff-coloured 'Butternut' fruit is generally straight but occasionally curved. The neck is entirely usable because the small seed cavity is confined to the bulbous base of the fruit. 'Waltham Butternut' is similar, but produces a higher proportion of fruit with straight necks. It was obtained by crossing 'New Hampshire Butternut' with an African plant introduced into the USA and has been a very popular cultivar ever since its commercial introduction in 1970. 'Ultra' has large fruit with an extended neck and is sometimes used by processors.

'Cheese', one of the oldest *C. moschata* cultivars grown in the USA, has a flattened, ribbed fruit similar in shape to a cheese box, with buff-coloured rind and deep orange flesh. It was popular for canning and stock feed during the 19th century.

After *C. pepo*, *C. maxima* exhibits the greatest diversity of fruit types. As with the horticultural classification of *C. moschata*, not all of the localized landraces of *C. maxima* that have evolved can be placed in the following informal classification scheme, which is based on Ferriol and Picó (2008):

1. Banana. Fruit long, pointed at both ends, with a soft rind and brown seeds.
2. Delicious. Fruit turbinate, shallowly ribbed, with a hard rind and white seeds.
3. Hubbard. Fruit oval, tapering to curved necks at both ends, with a very hard rind and white seeds.
4. Marrow. Fruit oval to pyriform, tapering quickly at the apex and gradually towards the base, with white seeds.
5. Show. Fruit large, orange, with a soft rind and white seeds.
6. Turban. Fruit turban-shaped as a result of fruit tissue at the blossom end not covered with receptacle tissue.
7. Kabocha and Buttercup. Medium-sized turbinate fruit with small to medium exposed ovary. Soft rind and dense fine-textured flesh.

Some cultivars, such as the heavily warted 'Victor', were produced by hybridizing cultivars of different groups. The parentage of 'Victor' is believed to include hubbard and turban squashes (Tapley *et al.*, 1937).

'Buttercup' is a high-quality winter squash. Its small, dark green fruit has a 'button' – a protuberance at the blossom end where the mature ovary is not covered by the receptacle. This turban-group cultivar was bred by A.F. Yeager and released in 1931. 'Buttercup' plants, like those of most cultivars of *C. maxima*, are large vines. Yeager later crossed 'Buttercup' with a bush-habit USDA plant introduction of *C. maxima* and selected for bush plants with 'Buttercup'-type fruit set close to the crown. The best of these selections was named 'Bush Buttercup'. The Japanese kabocha types are similar to the buttercup group, but have flesh with a slightly greenish cast and a distinctive flavour. Buttercup and kabocha squash have some of the best culinary quality of the *C. maxima* winter squash types.

'Turk's Turban' is a turban squash with very colourful fruit, which are used for decoration. It is consumed as a winter squash. 'Hubbard' produces a large, oval fruit that stores well and is of good quality. It was introduced by James Gregory of Marblehead, Massachusetts in 1856, but was probably originally brought to New England from South America in the 18th century. Cultivar selections have been made that produce orange, green or grey fruit. 'Queensland

Blue' is a high-quality Australian cultivar in the 'Jarradale' group. The flattened, deeply ribbed fruit have a bluish-grey rind.

Show pumpkins are grown for forage in India, where there is much diversity. In some countries, cultivars producing massive orange fruit, such as 'Atlantic Giant' and 'Big Max', are cultivated for entry into contests for large fruit. Using special breeding stock of 'Dill's Atlantic Giant', growers with the champion fruit now routinely achieve weights of over 2000 lb (907 kg). In 2016, the record weight was achieved in Europe with a pumpkin that weighed 2625 lb (1191 kg). A combination of genetics for size and cultural practices is used to achieve these weights. To be considered a giant pumpkin (and not a giant squash) the fruit must be at least 80% orange or red.

Many squash cultivars were described and illustrated by Tapley *et al.* (1937). Some of these older cultivars still exist today. 'Hubbard', for example, has been grown for over 150 years and 'Butternut' has been popular for more than 50 years. Within each of these varietal types, many different cultivars now exist. The situation is less stable for *C. pepo*, where there has been a rapid turnover in popularity of F_1 hybrid summer squash cultivars. However, some older *C. pepo* cultivars, such as 'Acorn', 'Scallop' and 'Connecticut Field', all more than 100 years old, are still grown today. Descriptions of many recent hybrids and cultivar selections can be found in Facciola (1990). For a more comprehensive list of contemporary cultivars, consult commercial seed catalogues.

A major and relatively recent focus of many squash breeding programmes is disease resistance. Relative to the other major cucurbit crops, squash is very vulnerable to attack by viruses, fungi and bacteria. In addition to traditional plant-breeding methods, including interspecific hybridizations to transfer disease resistance alleles from wild to cultivated species of *Cucurbita*, virus coat protein genes have been incorporated to provide additional resistances (for more details, see Chapter 3). Seminis (a subsidiary of Bayer) has released several summer squashes (*C. pepo*) engineered with resistance to watermelon mosaic virus (WMV) and zucchini yellow mosaic virus (ZYMV) using the virus coat protein. Cultivars carrying this trait are usually designated with a 'II' or 'III' after the cultivar name.

Breeders continue to improve flesh quality in winter squashes, including deeper orange coloration and pro-vitamin A carotene content. Selection is also made for flavour, texture, consistency and freedom from bitterness.

Oved Shifriss transferred the gene for bicolour fruit (*B*) from a *C. pepo* ornamental gourd to a summer squash. In the gourd genetic background, the allele produces bicoloured fruit, each fruit on a plant having a variable proportion of green and yellow rind colour. By selection for *B* and the proper modifiers, it has been possible to breed summer squash cultivars with entirely yellow fruit that are high in vitamin A content compared with most summer squash. Symptoms of WMV may be masked on these cultivars; the infected fruit do not develop the unsightly green rings produced by the virus on other yellow-fruited squashes and consequently they can still be marketed. 'MultiPik' and other hybrid cultivars with *B* have become commercially important.

Although bush cultivars of *C. pepo* (e.g. most summer squashes) and *C. maxima* (e.g. 'Emerald') have been cultivated for many years, few *C. moschata* cultivars with short internodes, such as 'Burpee Butter Bush', have been available until recently. 'Burpee Butter Bush' provided the source of the semi-bush habit found in cultivars 'El Dorado' and 'La Estrella', released by the University of Florida in 2002, and 'Taína Dorada', released by the University of Puerto Rico Experiment Station in 2013.

WATERMELON AND RELATIVES

Introduction

Watermelon, *Citrullus lanatus* (syn. *C. vulgaris* Schrad.), is an important crop in China, Africa, India, USA and other areas with a long, warm growing season. The plants are fairly drought resistant, flourishing on fertile sandy soils in hot, sunny, dry environments. Worldwide consumption of watermelon fruit and their seeds is greater than that of any other cucurbit (see Chapter 2, Table 2.1).

Classification

Seven species of *Citrullus* are currently recognized. *C. lanatus* (Thunb.) Matsum. & Nakai is the sweet or dessert watermelon. The closely related egusi watermelon is *C. mucosospermus* (Fursa) Fursa. The citron, tsamma or preserving melon is *C. amarus* Schrad. (formerly *C. lanatus* var. *citroides*). Another closely related species is *C. ecirrhosus* Cogn., the tendril-less melon. More distantly related species (in order) are *C. rehmii* De Winter, *C. colocynthis* (L.) Schrad. and *C. naudinianus*. Three of the *Citrullus* species, *C. ecirrhosus*, *C. rehmii* and *C. naudinianis*, are not cultivated. All species in the genus are cross-compatible with each other to varying degrees. *C. lanatus* and *C. ecirrhosus* are more closely related to each other than either is to the colocynth, *C. colocynthis* (Navot and Zamir, 1987). Related genera (in order) include *Lagenaria*, *Benincasa* (*Praecitrullus*), *Acanthosicyos* and *Cucumis*.

Dessert watermelon and egusi watermelon have been crossed to add traits from egusi, with success being higher when dessert watermelon was used as the female parent. Crosses of citron and dessert watermelon may result in progeny having preferential segregation (and reduced pollen fertility). That makes it difficult, although not impossible, to obtain new (non-parental) combinations in plant breeding programmes.

Botany

Watermelon differs from other economically important cucurbits in having pinnatifid leaves, although there is a mutant with non-lobed leaves (Fig. 4.21).

Fig. 4.21. Leaf shape of watermelon (*Citrullus lanatus*), showing normal (left) and non-lobed (right).

The hairy stems are thin, angular and grooved, bearing branched tendrils. Dwarf watermelon cultivars with reduced internode length have been bred, but most commercial cultivars are highly branched vines, measuring up to 10 m in length. The root system is relatively extensive, but shallow.

Watermelon cultivars are usually monoecious, but the accessions that are heirloom cultivars often have andromonoecious flowering. The solitary, light yellow flowers are less showy than those of many other cucurbits (Fig. 4.22). The large, round to oblong or cylindrical fruit measure as long as 60 cm. The 1–4 cm thick rind is hard but not durable; the exocarp is light to dark green, either solid-coloured, striped or marbled. The bland to sweet-tasting flesh is usually red or pink, but may be orange, salmon yellow, canary yellow, white or green in some cultivars or landraces (Fig. 4.23). Watermelons vary considerably in seed colour (e.g. black, brown, tan, white, red or green), as well as shape and size, and seed characteristics can aid in cultivar identification.

Origin and history

All *Citrullus* species originated in Africa, but *C. colocynthis* also grows wild in India. Citron (*C. amarus*) originated in southern and central Africa, dessert watermelon in northeast Africa, and egusi watermelon in west Africa (Fig. 4.24). Dessert watermelon grows wild in Sudan, as well as Egypt, Ethiopia and Kenya. The progenitor species of dessert watermelon may be related to *C. rehmii* or *C. colocynthis*.

Fig. 4.22.   Female watermelon flower.

Fig. 4.23.   Fruit flesh colours of watermelon (*Citrullus lanatus*), with (left to right): scarlet red, canary yellow, coral red, orange, and salmon yellow.

Watermelon has a long history of cultivation in Africa and the Middle East. It has been an important vegetable in Egypt for at least 4000 years. By the 10th century AD, the crop was being grown in China and southern Russia. Commercial production of black watermelon seeds has been ongoing in north-western China for over 200 years. Watermelon was introduced to the New

Fig. 4.24. Fruit of citron (*Citrullus amarus*), along with other cucurbits collected in the Republic of South Africa.

World by the Spanish in the 16th century and quickly became popular with Native Americans. In the USA, it is commercially grown primarily in Florida, Georgia, California, Texas, North Carolina, South Carolina and Indiana.

Uses

Watermelon fruit make a delicious and refreshing dessert, especially esteemed in hot weather. For centuries, watermelon and relatives such as citron have served as an important source of water in the Kalahari Desert and other arid areas of Africa. Watermelons are mostly eaten fresh, but in Africa they are also cooked. The rind may be pickled or candied. In the southern states of the former USSR, juice from watermelon fruit is made into a fermented drink or is boiled down to a heavy, sweet syrup.

Watermelon seeds are powdered and baked like bread in India. Roasted seeds are eaten in the Orient and the Middle East. Some Chinese cultivars used for this purpose have been bred to have very large seeds. However, big seeds can be objectionable for watermelons eaten as a dessert, and Japanese breeders have successfully selected for small seed size (Fig. 4.25).

Cultivars and breeding

Watermelons as large as 159 kg have been reported, but most cultivars produce fruit weighing 1–15 kg. Fruit sizes can be micro (< 1.4 kg), mini (1.4–3.7 kg), icebox (3.8–6.1 kg), small, sometimes called pee-wee (6.2–8.3 kg), medium (8.4–11.1 kg), large (11.2–14.5 kg) and giant (> 14.5 kg). The icebox cultivar

Fig. 4.25. (a) Seed size of watermelon (*Citrullus lanatus*), showing tomato seed size, short, medium and long seed size (left to right). (b) Seeds of watermelon (*C. lanatus*), showing large green seeds on canary yellow fruit flesh.

'Sugar Baby', released by M. Hardin in 1956, produces small dark-green spherical fruit. It is popular with home gardeners and is widely grown in Asia. Also, there are cultivars that will mature in production areas having short seasons.

Fusarium wilt resistance, obtained from *Citrullus amarus* (citron), was first bred into 'Conqueror'. Resistant cultivars with improved quality were subsequently developed and became more valuable commercially. 'Klondike R-7', a cultivar with resistance to Fusarium wilt, was introduced by the University of California at Davis in 1937 and remained popular for many years.

'Charleston Gray' has fruit with elongate shape, grey (light green) rind, light red flesh, crisp texture, and sweet (10 °Brix) flavourful taste. It was released in 1954 by the USDA in Charleston, South Carolina, and was a leading cultivar in the USA for decades. From Kansas State University (C.V. Hall), several cultivars were released and have been important because of their high-quality fruit and resistance to Fusarium wilt and anthracnose. These included round-fruited 'Crimson Sweet' (released in 1963), elongate-fruited 'Allsweet' (released in 1973) and small-fruited 'Petite Sweet' (released in 1970). From the University of Florida (J. Crall and G. Elmstrom), several cultivars were important because of their high-quality fruit, along with resistance to Fusarium wilt and anthracnose. These included elongate-fruited 'Jubilee' (released in 1963), round-fruited 'Dixielee' (released in 1979) and small-fruited 'Mickylee' (released in 1986).

'Charleston Gray' and 'Crimson Sweet' are still popular in countries of Asia and Africa. Local selections in tropical West Africa include 'Accra', 'Anokye' and 'Volta'. 'Arka Jyoti' is an Indian cultivar and 'Zhongyu No. 1' is a Chinese cultivar. Some accessions have names like tarmuj, but that is just one of the common names for watermelon in the various languages of India that include tarbuj, kallangadi, he'nd wend, thannir mathan, kalingad, tarvuja, tarbuja, darbusini and puchakayi.

Seedless watermelon cultivars were developed as a result of the discovery by Kihara (1951) that triploid watermelons are mostly seedless. Triploid hybrids are produced by crossing two inbred lines: a tetraploid female parent with a diploid male parent (the reciprocal cross is unsuccessful). Among the first successful cultivars (6 kg size) were 'Tri-X-313' from O.J. Eigsti (American SunMelon, now part of Syngenta) and 'AC 5244' from W. Barham (Barham Seeds, now part of Seminis). Later, mini seedless (1.4–3.7 kg) cultivars were developed. One of the first successful ones was 'Petite Perfection' from X.P. Zhang (Syngenta).

Tetraploids are produced by the application of colchicine or oryzalin to the seedlings of diploid lines and can be maintained by self- or sib-pollination. They are used as maternal parents in crosses with diploids. The resulting triploid F_1 hybrids, because of their odd chromosome number, are female and male sterile. Germinating pollen grains stimulate enlargement of the triploid ovaries. The fruit are not always entirely seedless, but may have small, empty seed coats and occasional hard seeds. Few tetraploids have been successful in producing triploid hybrids in the marketplace, either in the small (6.2–8.3 kg) or mini (1.4–3.7 kg) fruit sizes.

Because seedless cultivars are male sterile, growers must plant a pollenizer (diploid cultivar) in order to get fruit set. Usually, one row of the diploid is planted for every three or four rows of triploid in the production field. The diploid and triploid cultivars should be distinguishable, such as one having round fruit with narrow stripes (seedless) and the other elongate fruit with wide stripes (seeded), so that they can be separated at harvest. An adequate bee population (two active hives per hectare) is needed for successful fruit set and to minimize the percentage of cull fruit. A second method for pollenizer use is to plant the diploid in the triploid rows, with one diploid plant for every four triploid plants. The pollenizer may be a specialized cultivar with highly branched vines, many staminate (male) flowers and small fruit that are easily crushed when the harvest crew steps on them. The first cultivar of that type was 'SP-1' from X.P. Zhang at Syngenta.

Seed production of triploid cultivars is expensive and seeds germinate poorly if conditions are less than optimum. Therefore, growers often use transplants rather than direct-seeding the field. Some growers remove the seed coats before planting, in order to improve germination. Production of triploid seeds and transplants, along with the need for pollenizers, makes seedless cultivars significantly more expensive than seeded cultivars. However, they have become popular in North America, Europe, Japan and Africa.

In addition to seedlessness and disease resistance, breeders have selected for earliness, high yield, improved flesh characteristics such as higher sugar content, and thick or thin but tough and flexible rind, the latter to reduce damage during shipment. Dwarf vined (bush type) cultivars have been developed and are useful for home gardeners having limited space, but the commercially important cultivars are tall (vining type).

Commonly used genes for cultivar development are tough rind (*EE*), wide stripes ($g^W g^W$), elongate (*OO*), oval (*Oo*) or round (*oo*) fruit shape, scarlet red ($Y^{Scr}Y^{Scr}$) or coral red ($Y^{Crl}Y^{Crl}$) flesh colour, black (*DD RR TT WW*) seed colour, and short seed size (*ll ss*). There are also genes for disease resistance, as follows. Anthracnose resistance is from *Ar-1* for race 1 and *Ar-2-1* for race 2 resistance. Alleles *db* and *Fo-1* provide resistance to gummy stem blight and race 1 of Fusarium wilt, respectively. However, gummy stem blight resistance appears to be due to more than just a single gene. Susceptibility to powdery mildew is governed by *pm*. Resistance to the watermelon strain of papaya ringspot virus is controlled by a single recessive gene, *prv*. A moderate level of resistance to zucchini yellow mosaic virus is conferred by a single recessive gene *zym-FL*. A high level of resistance to the Florida strain of ZYMV is controlled by a single recessive gene, *zym-FL-2*; not the same as *zym-FL*. Resistance to the China strain of ZYMV is controlled by a single recessive gene *zym-CH*. Finally, a single dominant allele, *Zym*, confers resistance to zucchini yellow mosaic virus. In China, there is a large market for confectionery (edible seeded) type and plant breeders have selected for high yields of large seeds. They are also seeking to improve resistance in their landraces to Fusarium wilt, anthracnose and gummy stem blight (Zhang and Jiang, 1990).

Genetic diversity

This lack of diversity in general is in contrast to the large morphological variation observed in flesh colour, seed size and colour, and fruit rind pattern (discussed in detail in the above section) (Fig. 4.26). Watermelon, which originated

Fig. 4.26. Fruit shape and rind pattern of watermelon (*Citrullus lanatus*), showing: golden rind, spotted rind and intermittent stripes (first column); gray rind, narrow stripes, medium width stripes and wide stripes (second column); and light solid, medium solid and dark solid rind (third column).

in Africa at least 4000 years ago, was domesticated as a source of nutrients (seeds and flesh) and water in semi-arid and arid climates (Paris, 2015). From Africa, the dessert watermelon was brought to the Mediterranean in the 2nd century AD, reaching Italy and France in the 5th and 6th centuries. The crop was continually improved by agriculturalists as it became more widely used.

From its popularity in the Mediterranean, dessert watermelon spread to Europe and then China (secondary centre of diversity) through the Silk Road (McCreight *et al.*, 2013). The tertiary centre of diversity is central Asia. India was thought to be a secondary centre for watermelon due to the presence of tinda there. However, tinda (formerly *Citrullus*) has since been classified as *Praecitrullus*. India was probably introduced to watermelon during the Islamic conquests in the medieval period.

The fruit rind of preserving melon or citron (*C. amarus*, formerly *C. lanatus* var. *citroides*) is used to make pickles or conserves and the fruit are also fed to livestock. Citron has hard, white or pale green flesh that is bland to bitter. The large seeds are of various colours, including light green (e.g. 'Colorado Preserving Melon'). In commercial production, over 100 fruit per plant and up to 200 seeds per fruit give seed yields of 500–700 kg ha^{-1}.

Citron grows wild in Africa and as an escape elsewhere. It is a weed in watermelon-growing areas of North America, causing severe problems in cotton and sorghum fields in Texas, where three distinct citron types exist. Citron crosses readily with watermelon, so seed production fields should be isolated and monitored.

Egusi watermelon (*Citrullus mucosospermus*)

Egusi watermelon (*C. mucosospermus* (Fursa) Fursa) is the most closely related species to dessert watermelon. Egusi watermelons, indigenous to western Africa, are usually spherical with bitter flesh and have a soft, mucilaginous seed coat. They are cultivated for the consumption of their seeds, which range in size and shape.

Colocynth (*Citrullus colocynthis*)

The colocynth, *C. colocynthis*, is a perennial that is sparingly cultivated for medicinal use of the dry fruit pulp of unripe but mature-sized fruit. Although colocynth fruit have bitter white flesh, the non-bitter seeds are eaten and used for cooking oil in Africa. This species grows wild in northern Africa and south-western Asia and exists as an escape from cultivation in other areas, including Australia and southern Europe.

MINOR CROPS

Major cucurbit species, extensively cultivated globally, include cucumber, watermelon, melon, pumpkin and squash (discussed in Chapter 4). In addition to those already discussed in the previous chapter, there are many other cucurbits cultivated for food and other uses (see Chapter 1, Table 1.1). These cucurbitaceous vegetables, often referred to as minor cucurbits (due to non-global cultivation and consumption), include bitter gourd or bitter melon (*Momordica charantia* L.), spiny gourd (*Momordica cochinchinensis* Spreng.), balsam apple (*Momordica balsamina* L.), teasel gourd (*Momordica dioica* Roxb.), bottle gourd (*Lagenaria siceraria* (Molina) Standley), wax gourd (*Benincasa hispida* (Thunb.) Cogn.), sponge gourd (*Luffa cylindrica* Roem., syn. *Luffa aegyptiaca* Mill.), ridge gourd (*Luffa acutangula* (L.) Roxb)., snake gourd (*Trichosanthes cucumerina* L.), pointed gourd (*T. dioica*), chayote (*Sechium edule* (Jacq.) Sw)., ivy gourd (*Coccinia grandis* (L.) Voigt) and round melon (*Praecitrullus fistulosus* (Stocks) Pangalo). These cucurbits are primarily cultivated by smallholder farmers in Asia and are often part of home, school and community gardens there. Crops in this chapter are in order by taxonomic name (Latin binomial), since many of the crops have no widely accepted common name.

Benincasa hispida (Wax Gourd, Winter Melon)

Wax gourd (*Benincasa hispida* (Thunb.) Cign.) is also known by many other names, such as ash gourd, white gourd, white pumpkin, winter melon, tallow gourd, Chinese preserving melon and Chinese squash. It is a monotypic genus in the Cucurbitaceae family. The name wax gourd refers to the thick, waxy cuticle that typically develops on mature fruit. High temperature favours formation of the waxy bloom. It is an important vegetable in China, India, Bangladesh, Philippines, Thailand, Indonesia, Vietnam, Turkey and Iraq. The Indo-China region is considered as the centre of origin.

It is an annual climber or creeping vine up to 6 m long, with branched tendrils (bifid or trifid) up to 35 cm long. The large hairy leaves are lobed and have long petioles. Wax gourd is monoecious with solitary male and female flowers. Various fruit shapes, sizes and colours are available in wax gourd (Fig. 5.1). Fruit are round, oval, oblong, long cylindrical or short cylindrical. Fruit skin colour ranges from light green to dark green and speckled green. Fruit may bear strong, medium or weak wax and some are waxless. Mature fruit can attain a length of 80 cm and diameter of 35 cm, weighing up to 50 kg, depending upon the cultivar (Fig. 5.2). Indian consumers prefer round or oblong fruit (6–8 kg) of light green to dark green colour, whereas Chinese consumers prefer long cylindrical, heavy fruit with dark green colour. Long cylindrical fruit (1–2 kg) with dark green colour and white specks are preferred by consumers in Vietnam. Consumers in Thailand prefer short cylindrical fruit (2–4 kg) with light green to dark green colour.

Fig. 5.1. Wax gourd: variation of fruit size, shape, colour and skin pattern among cultivars.

Fig. 5.2. Wax gourds (winter melons) in storage at Ayers Creek Farm in Oregon. (Photo: J.R. Myers.)

Wax gourd cultivars popular with Indian growers include 'MAH-1 F1' (Mahyco), 'No 600 F1' and 'No 700 F1' (Sungro Seeds), 'Heera F1' and 'Greena F1' (Chia Tai Seed), 'Sowmya F1' (Bejo Sheetal), 'Pearl F1', 'Jade F1' and 'Gold 195 F1', a waxless cultivar (East-West Seed), 'Virat *F1*' (tolerant to begomoviruses) and 'Siddhi F1' (Rasi Seeds), 'Kashi Ujwal', 'Kashi Dhawal', 'Kashi Surabhi', 'Indu' (tolerant to begomoviruses) and 'KAU local'. The Field Crops Research Institute (FCRI) in Vietnam is involved in wax gourd breeding. Improved cultivars released by that institute include 'Wax gourd No. 1', 'Wax gourd No. 2', 'Wax gourd Thien Thanh 5', 'Wax gourd Sac', 'Wax gourd Chu Thap' and 'Wax gourd Da'.

Both mature and immature fruit are consumed. They may be eaten raw, but more often cooked or pickled. Immature fruit are prepared like summer squash, whereas chunks of the mature fruit are used to make soup, which is sometimes canned. In China, mature wax gourds are sold by the slice. Occasionally, the entire mature fruit is steamed, often stuffed with lotus seeds, vegetables, meat or other ingredients. In India, the fruit is boiled in sugar syrup to make a confectionery known as 'petha'. Young leaves, vine tips and flower buds are boiled and eaten as greens. Seeds are consumed fried, but usually as a medication instead of food.

Coccinia grandis (Ivy Gourd)

Ivy gourd (*Coccinia grandis* (L.) Voigt) occurs wild and cultivated from Africa to Southeastern Asia. It is commonly called 'kundru' in India. Ivy gourd is also grown and has escaped in Latin America, Australia and other tropical areas. This species is perennial and dioecious, with glabrous stems growing up to 20 m in length. Flowers are white in Indian cultivars, and white or yellow in African landraces (Jeffrey, 1967). The fleshy, ellipsoid fruit measure 6–10 cm long. They are green, often with white stripes, when young, turning scarlet at maturity (Fig. 5.3). Landraces with fruit that are non-striped are also available. Mature fruit are consumed raw, cooked or candied. Leaves, young shoots and immature fruit are cooked as vegetables. Various plant parts are used medicinally. The plant is sometimes grown as an ornamental on arbours.

Ivy gourd is usually propagated by cuttings in order to have ten times as many female as male plants in the field. Stem cuttings are placed 1–2 m apart in each row under the trellis. Irrigation may be required and the application of nutrients is desirable. Fruit of some landraces can develop parthenocarpically. A single plant may bear over 300 fruit in a season. Cultivars and landraces that are popular with farmers in India are 'Indira Kundru-5', 'Indira Kundru-35', 'Sulabha', 'DRC-1' and 'DRC-2' and 'Co-1'. The wild germplasm of ivy gourd that originated from northeastern India is genetically divergent from commercial cultivars popular in India (Singh and Dathan, 1990).

Fig. 5.3. Ivy gourd: fruit variability among Indian cultivars (Photo credit: Dr. L.K. Bharathi.)

Cucumis anguria (West Indian Gherkin)

Cucumis anguria var. *anguria* is known as burr gherkin or West Indian gherkin. The term gherkin is imprecise, since it has been used both for *C. anguria* and for pickling type cucumber.

West Indian gherkin is a fast-growing annual that is cultivated in Latin America and the West Indies. It is also listed in seed catalogues and occasionally grown by home gardeners in the USA and other countries. In many of these areas, West Indian gherkin has become a persistent weed.

C. anguria was previously considered to be the only species in the genus native to the Western Hemisphere, and its common name reflects its presumed origin in the Caribbean. Actually, this cultigen was introduced to the West Indies and Brazil from Africa by the slave trade. The African taxon once labelled as *C. longipes* Hook. f. and now reclassified as *C. anguria* var. *longaculeatus* Kirkbride is considered the progenitor of domesticated var. *anguria*. A salient step in the domestication of this species was selection for non-bitter fruit, which was accomplished by a single-gene mutation.

The flowers, tendrils and lobed leaves of the West Indian gherkin are smaller than those of cucumber. The oval fruit are light green to pale yellow and covered with short fleshy spines. Although the fruit are smaller (5 cm diameter) than most cucumbers, the peduncle reaches up to about 20 cm in length (Baird and Thieret, 1988), which can be longer than cucumber fruit. The plants produce great quantities of fruit throughout the growing season. Immature fruit are eaten fresh, pickled or cooked in curries, soups or stews.

Cucumis metuliferus (Horned Melon)

This species is also known as African horned cucumber, because it is from Africa and its foliage resembles that of cucumber. However, it is not at all closely related to cucumber, but is closer in phylogeny to melon and other species of the subgenus *Melo*. It is also known as horned melon or 'kiwano'.

The oblong fruit are 10–15 cm long and have broad-based, stout spines measuring ca 1 cm in length. After about a month on the vine, the green, mottled fruit reach maximum weight, then take another 2 weeks to reach maturity, turning sweeter and orange in the process. They are quite colourful, with their bright orange skin, deep green flesh and bizarre spines (Fig. 5.4). They have

Fig. 5.4. African horned melon. (Image by jasper-m from Pixabay.)

been said to have a flavour combining that of bananas, lemons and passion fruit, but others consider this hyperbolic and to them the fruit has an insipid or astringent taste. Fruit from indigenous wild plants are bitter. The leaves are sometimes cooked and eaten in Africa.

In recent years, New Zealand growers, encouraged by their success in marketing kiwi fruit, have exported African horned melon under the trade name 'kiwano'. The fruit have a long storage life and ship well. The commercial potential of this crop increased as researchers improved seed germination, stabilized fruit characteristics such as size and uniform coloration, enhanced flavour, and extended storage life under humid conditions (Benzioni *et al.*, 1993).

Similar to other cultivated species of *Cucumis*, the African horned melon can become naturalized in areas where it is not native. For example, it has been a pest in sugarcane fields and on farms in Queensland, Australia since the 1930s.

Cucumis setosus

Cucumis setosus Cogn. has fruit that resemble cucumber, but smaller. The leaves also resemble those of cucumber. It has been used directly as a crop, in the same way as cucumber.

However, *C. setosus* is not crossable with *C. sativus*, but only with *C. melo*. It has 12 pairs of chromosomes (John *et al.*, 2014), like melon, and may be considered a source of genetic diversity for melon improvement.

Cucurbita foetidissima (Buffalo Gourd)

Attempts have been made to domesticate buffalo gourd (*C. foetidissima*). This xerophytic perennial is native to the arid deserts of southwestern USA and Mexico. Its ability to withstand drought has attracted interest in its commercialization. The fruit are 7–10 cm in diameter and weigh 120–150 g, each containing 300 seeds. Leaves are heart-shaped. Plants form a taproot for storage and overwintering.

Buffalo gourd produces a high yield of seeds that are rich in protein and oil. The foliage can be used for cattle feed and the large tuberous roots are a source of carbohydrates. F_1 hybrids of buffalo gourd have been bred using a gynoecious accession of the species as the female parent and a monoecious inbred line as the male parent. Despite previous research, however, this species is not yet being grown commercially.

Cyclanthera pedata var. *edulis* (Stuffing Cucumber)

Cyclanthera pedata var. *edulis* Schrad. is native to the Americas, where it often occurs as an escape. Stuffing cucumber has also long been cultivated in Asia, sometimes at elevations of up to 2000 m because of its cold tolerance.

The glabrous vines have a strong odour when crushed, somewhat like a cucumber. Leaves are large and deeply palmately lobed. The species is mon-oecious, with staminate flowers in racemes and solitary pistillate flowers. The green or white flowers are relatively small for a cucurbit. The puffy, partially hollow fruit is like a bladder. It is yellowish green, measures 5–15 cm long and has soft spines, a tapered neck and black seeds.

The foliage is eaten fresh or cooked. Fruit are eaten raw as a substitute for cucumber, or cooked, often after removing the seeds and stuffing the fruit with various ingredients. Seeds are edible as well. Fruit can be harvested throughout most of the growing season.

Lagenaria siceraria (Bottle Gourd)

Bottle gourd (*Lagenaria siceraria* (Mol.) Standl.), also known as calabash and white-flowered gourd, is native to Africa (Richardson, 1972). Its African origin is confirmed by the recent nuclear ribosomal DNA (nrDNA) analysis of African and Asian cultivars (N'dri *et al.*, 2016). Archaeological studies in-dicate its appearance in Asia and the Americas 7000–10,000 years ago by means of human migration or ocean currents (Erickson *et al.*, 2005; Kistler *et al.*, 2014). Collections of Indian origin are genetically distinct from the col-lections made in Africa, the Americas and Turkey (Levi *et al.*, 2009; Gürcan *et al.*, 2015). The *Lagenaria* genome is one of the most recent cucurbits to be se-quenced. *Lagenaria* diverged from *Citrullus* 10.4–14.6 million years ago, from *Cucumis* 17.3–24.3 million years ago and from *Momordica* 29.2–41.0 million years ago.

Immature (tender) fruit are peeled and then cooked (fried) or used in curries. Fresh juice is made from the fruit in India for its cooling, diuretic, anti-bilious and pectoral properties (Minocha, 2015). Heiser (1979) gave an interesting account of various uses of the mature dried shells of bottle gourd. Those include cups, milk pails, ladles, fishing floats, barrels, penis sheaths, carvings and musical instruments. Bottle gourd fruits and seeds can come in unusual shapes, making them easily recognizable (Fig. 5.5). Market segments for bottle gourd are based on fruit shapes, including long and cylindrical, globular, round, crooked neck, pyriform and elongate (Fig. 5.6).

Bottle gourd is chiefly suited to sunny, semi-dry, low-elevation areas, but can be grown in wet tropical climates as long as soils are well drained. Production temperatures can be 18–30°C. In temperate countries, some bottle gourd

Fig. 5.5. (a) Bottle gourd seeds, measuring 15–17 mm long. (b) 'Maranka', an unusual cultivar of bottle gourd. This fruit is about 30 cm long. (Photo credit: Ravishankar Manickam.)

Fig. 5.6. Bottle gourd: variation of fruit size, shape and colour among cultivars and landraces (Photo credit: Dr. Khushwinder Kaur Dhillon, Hira Singh.)

cultivars can be grown outdoors, but others may need season extension aids such as production in unheated high tunnels to fully mature fruit with thick and durable rinds. Trellising is necessary. Immature fruit are harvested 60–90 days after sowing. Fungal diseases such as powdery mildew, downy mildew and anthracnose, along with virus diseases, are major production constraints. Young fruit can be covered with mosquito netting to keep insects from laying eggs in the rind.

Important bottle gourd cultivars developed by private seed industries in India include 'Vidya' and 'Swati' (Sungro Seeds), 'Warad' (Mahyco Seeds), 'Anmol' and 'Gadda' (East-West Seed), 'Sharda' (Seminis), 'Anokhi' (Bayer India) and 'Mallika' (Bio Seeds). 'Narendra Shishir', a cultivar from Indian Agricultural University, has resistance to anthracnose, downy mildew and an unspecified viral disease complex (Singh, 2013). In Japan, South Korea and China, bottle gourd is extensively used as a rootstock to control soil-borne diseases, particularly Fusarium wilt of watermelon as a majority of the rootstocks of bottle gourd are non-hosts to *Fusarium oxysporum* f. sp. *niveum*, the causal organism of Fusarium wilt of watermelon (Cohen *et al.*, 2007; Bruton *et al.*, 2009). The bottle gourd accession PI 271353 from India was resistant to powdery mildew caused by *Podosphaera xanthii* (Kousik *et al.*, 2008). Two more breeding lines, USVL#1-8 and USVL#5-5, derived from Indian bottle gourd germplasm, were reported resistant to zucchini yellow mosaic virus (ZYMV), the watermelon strain of papaya ringspot virus (PRSV-W), watermelon mosaic virus (WMV) and squash vein yellowing virus (SqVYV) (Ling *et al.*, 2013). Commercial bottle gourd rootstocks 'FR-Strong', 'Emphasis', 'Marcis' and 'WMXP-3938' were reported resistant to crown rot caused by *Phytophthora capsici* (Kousik *et al.*, 2012). Salinity-tolerant rootstocks of bottle gourd have also been reported (Yetişir and Uygur, 2010).

Luffa spp. (Luffa Sponge Gourd, Ridge Gourd)

Genus *Luffa* (commonly called as 'loofah' in English) comprises two cultivated species: sponge gourd (*Luffa cylindrica* Roem., syn. *L. aegyptiaca* Mill) and ridge gourd (*L. acutangula* (L.) Roxb.), and both are clearly differentiated on the basis of molecular markers (Prakash *et al.*, 2014). *Luffa* is regarded as an Old World genus with four wild species: *L. graveolens* Roxb. (var. *longistyla*), *L. echinata* Roxb., *L. tuberosa* Roxb. and *L. umbellata* Roem (Seshasdri and More, 2009). Wild species of *Luffa* from the New World include *L. quinquefolia* (Hook and Arn) Seem. and *L. astorii*. Another wild species was reported from Australia's Northern Territory (Telford *et al.*, 2011). The Indo-Burma region and India are considered the centres of diversity of sponge gourd and ridge gourd, respectively (Whitaker and Davis, 1962).

Loofah vines are long, sometimes exceeding 10 m. The slightly hairy stem is five-angled. Sponge gourd has large shallowly to deeply lobed leaves, which are mottled with white silvery areas between the veins when young.

Those of ridge gourd are shallowly lobed to merely angled and lack the mottling. Tendrils are usually three- to five-branched. Most *Luffa* species are monoecious, but *L. echinata* is dioecious. The yellow flowers are relatively large and conspicuous. Female flowers are solitary, often occurring in the same leaf axil as the long-stalked raceme of male flowers. Flowers of sponge gourd and ridge gourd open in early morning and late afternoon, respectively. Ridge gourd flowers are about half of the size of sponge gourd flowers. The cylindrical to club-shaped fruit of the domesticated cultivars are mostly non-bitter and measure 15–60 cm or longer. There is great variability in *Luffa* for fruit shape, size and colour (Fig. 5.7). Fruit of sponge gourd are smooth, whereas those of ridge gourd have distinct longitudinal ribs. Fruit are light or bright green when immature, turning yellow to brown when mature. Mature fruit dehisce at the blossom end, enabling the black, brown or rarely white seeds to disperse from the dry fibrous endocarp.

Immature fruit are eaten as vegetables, boiled, peeled and fried and used in curries. Mature fibrous mesocarp of sponge gourd fruit is used as scrubbing sponges, after removing the skin (exocarp) and shaking out the seeds (Fig. 5.8). The USA imports millions of them from Asia each year. Korea, China, Guatemala and Colombia are major producers of loofah sponges.

Fig. 5.7. *Luffa* spp: fruit variation for size, shape and colour among cultivars of *L. cylindrica* (a–i) and *L. acutangula* (j–n). (Photo credit: Dr. Prashant Kumar.)

Fig. 5.8.   Fibrous spongy interior of smooth luffa.

Most of *Luffa* breeding research is concentrated in India. Commercial cultivars of sponge gourd popular with Indian farmers include 'Harita' (Mahyco), 'NHSG' (Nirmal Seeds), 'NS 441' (Namdhari Seeds), 'Maya' (Bio Seeds), 'Nutan' and 'Sunali' (Sungro Seeds), 'Lohit' (Tropica Seeds) and 'Alok' (VNR Seeds). Popular hybrid cultivars of ridge gourd include 'MHRG-7' and 'Surekha' (Mahyco), 'Aneeta' (Advanta India), 'Gaurav' and 'Pallavi' (Sungro Seeds), 'NS-3' (Namdhari Seeds), 'Naga', 'Mallika' and 'Rama' (East-West Seed). Popular sponge gourd open-pollinated cultivars derived through selection from local landraces, released by public institutes in India include 'Pusa Chikni', 'Pusa Sneha', 'Pusa Supriya', 'Pant Chikni Tori-1' and 'Azad Tori-2'.

Tomato leaf curl New Delhi virus (ToLCNDV) is a serious virus disease infecting *Luffa* cultivation in Asia. Sponge gourd cultivar 'DSG-6', a selection from a landrace originated from West Bengal (India), is reported resistant to ToLCNDV (Islam *et al.*, 2011; Munshi *et al.*, 2015). The resistance is controlled by a single dominant gene and molecular markers linked to this gene have been developed (Islam *et al.*, 2010). Ridge gourd cultivar 'Arti' (VNR Seeds) is the first begomovirus-resistant hybrid released in India. In India, a lesser-known hermaphrodite local cultivar 'Satputia' (sometimes classified as a separate species, *L. hermaphrodita*) is commonly cultivated in Indo-Gangetic plains of Bihar and Uttar Pradesh.

Momordica angustisepala and *M. cymbalaria*

Momordica angustisepala (sponge plant), a native of Africa, is sometimes cultivated in Ghana. The thick stems are pounded and worked to yield white absorbent fibres that are used in washing. *M. cymbalaria* (syn. *M. tuberosa* (Roxb.) Cogn.) ranges from northeastern Africa to India. It is cultivated occasionally in India, where fruit provide food and medication.

Momordica balsamina (Balsam Apple)

Balsam apple (*Momordica balsamina* L.) is widely distributed from South Africa northwards to tropical East and West Africa and is also found in India and Australia. It is sometimes confused with wild plants of bitter gourd, which have similarly sized fruit. However, the mature fruit of balsam apple is usually beaked (Fig. 5.9) and not as spiny as bitter gourd. Fruit are spindle-shaped, green, with six to nine regular or irregular rows of cream or yellowish blunt spines. Balsam apple also differs from bitter gourd by having relatively smooth seeds, very slender short stems, leaves that are less deeply lobed and shorter staminate pedicels. The leaves and young fruit are cooked and consumed as a vegetable in Africa. It is widely used there as a medicinal plant. *M. balsamina* is genetically diverse from the genepool of cultivated bitter gourd (*M. charantia*) (Dhillon *et al.*, 2016a).

Momordica charantia (Bitter Gourd)

The genus *Momordica* contains about 45 species native to the palaeotropics, chiefly Africa. The name for the genus is derived from the Latin word 'mordicus', meaning bitten. It refers to the fact that the jagged edges of the seed look like bite marks. At least six species are under cultivation, and others are variously used. Two species, *M. charantia* and *M. balsamina*, have escaped from cultivation and become naturalized in the neotropics. Weedy *M. charantia* is particularly widespread and a nuisance in agricultural areas such as citrus groves in Florida. These plants can flower and fruit throughout the year as long as temperatures remain high.

Bitter gourd (*Momordica charantia* L.), also known as bitter melon, is an important cucurbitaceous cash crop in Asia, where it is cultivated on approximately 340,000 ha annually (McCreight *et al.*, 2013). The crop is grown in some African countries such as Ghana, Zambia, Congo and Madagascar, and fresh fruit is exported to Europe and the Middle East to supply the fresh market demand by expatriate Asian communities. The crop is also cultivated on a smaller scale in southern USA and parts of Australia (Northern Territory,

Fig. 5.9. Balsam apple fruits.

Queensland, New South Wales, Victoria) where mainly Asian hybrid cultivars are cultivated for local consumption by Asian communities (Morgan and
Midmore, 2002). Current biogeographical analyses indicate that *M. charantia*
originated in Africa (Schaefer and Renner, 2011) and probably was domesticated in eastern India and southern China (Reyes *et al.*, 1994).

Bitter gourd fruit are consumed in various ways such as fresh sliced salad,
boiled, steamed, fried and curried. Bitter gourd recipes are documented by the
World Vegetable Center in Taiwan (avrdc.org). The bitter taste of fruit is caused
by saponins, momordicosides K and L as well as momordicines I and II (Yasuda
et al., 1984; Harinantenaina *et al.*, 2006). The fruit can be parboiled or soaked
in saltwater before cooking to reduce the bitterness.

The species is monoecious (bears separate male and female flowers on
the same plant) and cross-pollinated. Bitter gourd is available in tropical
Asian vegetable markets throughout the year. Consumers prefer bitter gourd
fruit at the immature or unripe stage and farmers harvest fruit for marketing
about 12–20 days after fruit set (generally 60–70 days from sowing), depending on the cultivar and temperature. Immature fruit have a bright appearance and the immature seed coat is creamy white. Mature fruit have
yellow flesh with red arils covering the seed and usually split, rendering them
unmarketable. Consumer preferences for fruit colour, shape, size and skin
pattern vary among and within countries. For example, South Asian consumers prefer small to medium-size spindle-shaped and spiny fruit, whereas
long cylindrical and smooth fruit are preferred by consumers in Thailand.
Highly bitter fruit are preferred by consumers in South Asian countries,
whereas less bitter fruit are in demand by consumers in Southeast Asia
(e.g. Thailand and Vietnam). White fruit are used in soups in Taiwan and
India. Fruit colour can be white, cream, light green, medium green, or dark
green. Fruit shape can be spindle (with or without narrow neck and white
blossom end), conical, elliptical, or cylindrical. Fruit develop irregular longitudinal ridges and warty skin, depending on the cultivar. Based on these fruit
traits, nearly 20 market types of bitter gourd exist in Asia and nearly half of
those types are cultivated in India, China, Nepal, Bangladesh, Myanmar and
Sri Lanka (Fig. 5.10).

Bitter gourd is adapted to the tropics and subtropics and grows best in
well-drained sandy loam or silt loam soils high in organic matter with a pH
of 6.0–6.7 and an air temperature of 24–27°C (Desai and Musmade, 1998).
Bitter gourd vines are supported on high fences or overhead trellises, making it
a labour-intensive crop. Bitter gourd breeders have been effective in improving
the crop in the past two decades for fruit types that fit particular market niches.
Breeders have focused mainly on elite crosses to permit them to retain the gains
made in previous years. However, this has led to low genetic diversity among
bitter gourd cultivars over various market segments, a situation confirmed by
analysis using simple sequence repeat (SSR) markers at the World Vegetable
Center (Dhillon *et al.*, 2016a,b). This study further suggested that bitter gourd

Fig. 5.10. Bitter gourd: fruit variation within eight market types.

genetic resources originating from South Asia, Southeast Asia and East Asia are genetically divergent from each other.

Hybrid cultivars of bitter gourd are more popular with growers in Asia, and most of the commercial breeding activities are pursued by seed companies. The seed market for bitter gourd in Asia is valued at 16 million euros. In India, the bitter gourd seed market size is 900 t (250 t hybrid and 650 t open-pollinated). Hybrid cultivars of bitter gourd that are popular with Indian growers include 'VNR 28' (short segment) and 'VNR 32' (VNR Seeds), 'Palee' (long segment) and 'Prachi' (East-West Seed), 'Vivek F1' (Sungro Seeds), 'Vishesh' (Golden Seeds), 'Arjuna', 'Pallavi', 'Raja', 'Noor' and 'Parijit' (Rasi Seeds), 'US 1315' (short segment, US Agriseeds), 'CT 108' (Chia Tai Seed), 'Abhishek' (Seminis) and 'FIGO' (Noble Seeds). Landraces still popular with Indian farmers include Jhalari, Faizabadi Karela, White Long, Green Long, Katai, and Jaipur Long. Faisalabad Long is a popular landrace in Pakistan. Favoured bitter gourd landraces in Bangladesh include Gazkarela, Rampali and Ranipukar, and important hybrid cultivars include 'Tia F1' and 'Kakoli' (Lal Teer Seed), 'Bolder' (Metal Seed), 'Shyama' and 'BT-03' (ACI Seed), 'BRAC Hybrid 1' and 'BRAC Hybrid 2' (BRAC Seeds). Dominant bitter gourd cultivars in the Philippines include 'Mestisa F1', 'Galaxy F1', 'Galactica F1' and 'Bonito F1' (East-West Seed). The market for bitter gourd in Vietnam is occupied by a single market segment with cultivars having dominant fruit traits such as light green, slightly spindle-shaped with blunt spines and smooth ridges, and medium long (10–15 cm). Major cultivars in this market segment include 'Thuy Phi' (Known-You Seed), 'HN 126' (Vina Seed), 'Apolo-17' (An Phu Nong Seed), 'Galaxy B1' (Viet Nong Seed), 'Jupiter' (En Vang Seed) and 'Sumo 742' (Southern Seed Corporation). The vegetable seed industry is lacking in Vietnam, so these popular bitter gourd hybrids are outsourced from seed companies in Thailand and China. Sometimes, the same hybrid is marketed under different names. This is a common marketing practice of seed companies. Important bitter gourd cultivars in Thailand include 'Moddum', 'Phetrae' and 'Noree 204' (Chia Tai Seed), 'Kiew Yok 16', 'Dotcom F1' and 'Sanya F1' (East-West Seed). Popular bitter gourd landraces in China include Dading (cone-shaped fruit), Zhenzhu, and Laigua. Gynoecious inbred lines (DBGy-201, Gy263B, OHB61-5) are now available that have improved combining ability for high early and marketable yield in bitter gourd hybrids (Iwamoto *et al.*, 2009; Dey *et al.*, 2010).

Momordica cochinchinensis (Gac)

Gac (*Momordica cochinchinensis* (Lour.) Spreng.), also known as spiny gourd or sweet gourd, is a tropical cucurbit that is native to the Southeast Asian region from South China to northeastern Australia, including Thailand, Laos, Myanmar, Cambodia, Vietnam and eastern India, where it is widely used as a food and traditional medicine. It is known by different names in various Asian countries such as 'gac' in Vietnam, 'fak kao' in Thailand, 'mak kao' in Laos, 'moc niet tu' in China and 'bhat karela' in India. The species name *cochinchinensis* derives from the Cochinchina region in the northern part of Vietnam.

After removing the spiny skin, immature fruit are cooked as a vegetable or prepared in curries. The Vietnamese use the red seed pulp (aril) of gac in a dish called 'xoi gac', prepared by mixing gac seed and red pulp with cooked rice to give a red colour and distinct flavour. A 'functional drink' prepared from gac fruit has been popular in the Thai market recently (Fig. 5.11).

Fig. 5.11. Bottled gac fruit juice for commercial sale in Thailand.

Fig. 5.12. Gac fruits.

Gac is a perennial climber and one plant produces 60–70 fruit in a season. Fruit weigh 0.6–2 kg, are ovoid and covered with spiny tubercles 3–4 mm long. They turn orange or red at maturity. Embedded in the orange-red pulp are large (ca 4 cm across) flattened black seeds that are sculptured with jagged edges (Fig. 5.12).

Gac is dioecious (male and female flowers on separate plants). Dioecious plants are propagated by cuttings to enable the grower to have a high proportion of female plants. Female and male plants should be planted in a ratio of 9:1. Plants can be also raised from seed. Gac cultivation requires a relatively warm growing period of 8 months and a temperature range of 22–30°C and high relative humidity of 82–85% (Bharathi *et al.*, 2015). Thai commercial gac farmers grow 833 plants per hectare (133 plants per rai) with spacing of 3 m between rows and 4 m between vines. Vines are pruned in January and flowers appear in February. Fruit is harvested from April to August. The fruit yield is 7.5 t ha^{-1} in the first year and increases to 12.5 t ha^{-1} in subsequent years (Dhillon and Senapa, 2012). Vines will produce for up to 100 years.

In current local cultivars of gac, yellow mesocarp and red seed pulp account for 50% and 25% of fruit weight, respectively. Growers hope that breeders can develop gac cultivars with equal quantities of yellow and red fruit components – red pulp without seed would be ideal. Researchers at Khon Kaen University, Thailand, have started breeding work on this species. Currently, local landraces of gac are cultivated in Asia.

Momordica dioica (Kaksa, Teasel Gourd)

Momordica dioica (kaksa, teasel gourd) is native to southern tropical Asia, growing wild from India to China. It is a dioecious perennial with solitary yellow flowers. The small (3–8 cm long) fruit are sweet, non-bitter and have soft spines. Seeds are pale yellow.

Kaksa is cultivated in India, where the immature fruit are used in curries. Young shoots and leaves are cooked as a vegetable. The succulent roots, which

 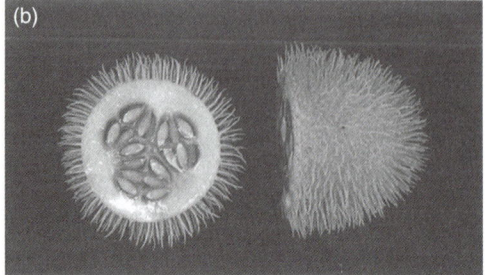

Fig. 5.13.   (a) Teasel gourd fruits. (Photo credit: Dr. Mohammed Abu Taher Masud.) (b) Teasel gourd measuring ca 4.5 cm in diameter at cross-section, including spines.

are larger in female plants, are also eaten. Because they are astringent and contain traces of alkaloids, the roots are used medicinally.

This species will not naturally cross with bitter gourd, which differs in sex expression, chromosome number and karyotype. However, Indian researchers have recently hybridized the two species and are beginning to select for desirable gene combinations.

Teasel gourd is occasionally grown for its ornamental fruit. The small, light green fruit is densely covered with fleshy spines measuring about 5 mm long (Fig. 5.13). This species has been called hedgehog gourd and teasel gourd for the resemblance of its fruit to a hedgehog and the spiny flower head of teasel, *Dipsacus fullonum* L., respectively. Teasel gourd has escaped from cultivation and become a feral weed in Mexico and other countries.

Praecitrullus fistulosus (Round Melon)

Round melon (*Praecitrullus fistulosus* (Stocks) Pangalo), commonly known as 'tinda' in India, is the only species in its genus. It was formerly classified as a botanical variety of watermelon, but it differs from watermelon in chromosome number and pollen characteristics and will not cross with the seven species of *Citrullus*. Based on electrophoretic pattern of seed storage proteins, this species has been reported to be genetically distant from all the *Citrullus* taxa (Singh and Matta, 2010).

Round melon is a popular summer vegetable in northern India, where it is believed to have originated. In India and Pakistan, the fruit are harvested when fully grown, tender, hairy and before the seed coats harden. The fruit is green in colour, spherical with a 5–7 cm diameter, depressed at each end, hispid when young and afterward glabrous (Fig. 5.14). The fruit are cooked as a fried vegetable. This species possesses a wide range of pharmacological properties, such as antioxidant activity in the fruit and antimicrobial activity in the

Fig. 5.14. Round melon fruits.

seeds (Sood *et al.*, 2012). There is not much genetic variability among cultivars grown in India and Pakistan. Popular cultivars of round melon in northern India include 'HT-10', 'Hisar Selection', 'Arka Tinda' and 'Punjab Tinda-1'. A single cultivar, 'Dilpasand', has been cultivated in Pakistan since the 1920s.

Sechium edule (Chayote)

Chayote (*Sechium edule* (Jacq.) Sw.), a single-seeded cucurbit, is also known as chow-chow, mirlitron and vegetable pear. The term chayote is derived from 'chayotli', its name in the ancient Aztec civilization of Mexico. *Sechium* was formerly regarded as monotypic, having only a single species in the genus, but now 11 species are recognized (Saade, 1996). Out of these, nine are wild species distributed in southern Mexico and two are cultivated: *S. tacaco* in Costa Rico and *S. edule* throughout the Americas and other parts of the world.

Chayote is native to Mexico, Guatemala and Costa Rica, where it has been cultivated since pre-Colombian times (Newstrom, 1991). It was introduced into India by the western missionaries (Singh *et al.*, 2015) and is popular with the farmers in Meghalaya, Sikkim and Mizoram states of northeastern India. From commercial fields in Mexico, Costa Rica, Brazil, Puerto Rico and Italy, chayote is shipped to markets in the USA, Canada and Europe. This crop has become an important vegetable in many tropical and subtropical areas of Asia, Africa, Australia and South America.

Chayote is a monoecious perennial and a vigorous grower, with vines reaching up to 20 m in length. The vine supports a single pistillate flower and a cluster of staminate flowers at the same node. Chayote fruit are white to dark green, pear-shaped to round, 7–20 cm long, often wrinkled and sometimes with a few spines. There is a considerable genetic variation in fruit with respect to size, shape, colour and spines (Fig. 5.15a). The single-seeded, fleshy fruit is

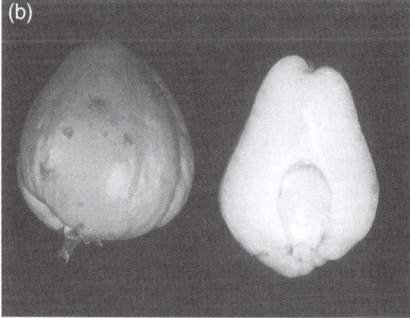

Fig. 5.15. (a) Chayote: fruit variability among landraces in Mexico. (Photo credit: Dr. Rafael Lira Saade.) (b) Chayote fruit with fleshy seed exposed.

viviparous, the developing seedling feeding off the pericarp. A fruit may even sprout while still on the vine. The white seed is large (up to 8 cm long) and flattened (Fig. 5.15b). A short photoperiod is required for chayote to flower. Since it may not flower until the autumn in temperate regions, a long growing season may be required. This crop is generally grown only in subtropical and tropical areas, where it may flower all year.

Fruit are eaten as cooked along with other vegetables, boiled, fried, or used in stews. Fruit flesh is also used to provide consistency to baby foods, juices, sauces and pastes. In India, the fruit and roots are also used as fodder for cattle.

There is ample genetic variation in chayote landraces and wild populations in Mexico and Central America that could be used for crop improvement. However, only two commercial types are being exported from Mexico, Costa Rica and Puerto Rico. The dominant type is a bland-tasting, obovoid, light green, smooth fruit that is actually intended for the food industry, but is also used as a vegetable. The second type is a small, round, white, smooth fruit. Two cultivars currently grown in Florida are 'Florida Green' and 'Monticello White'. Newstrom (1990) suggested that selections from other native landraces could produce a more palatable vegetable for export as well as providing the disease resistance that is lacking in the narrow gene base of commercial cultivars. Four heat-tolerant local landraces (Local-1, Local-2, Local-3, Local-4) were identified in Mizoram state of India (Singh *et al.*, 2015).

In the tropics, chayote can be cultivated at elevations of 800–2000 m. It thrives best in areas of fertile soils, high humidity (80–85%), ample well-distributed rainfall (1500–2500 mm) and mild temperatures (13–21°C) with relatively cool nights. Chayote is propagated by planting three to four whole fruit in a hill. It is also grown from cuttings to perpetuate desirable plants and avoid the genetic diversity of plants grown from seed. In India, Mizoram state is the biggest producer of chayote fruit and farmers train vines on horizontal trellises from which the fruit are supported. Plants on a trellis or other support should be widely spaced, at least 3 m apart, because of their larger size. They can be grown at a density of up to 1500 plants per hectare. Flowering occurs at a daylength of 12.5 h or less. Fruit require pollinators, but parthenocarpic fruit can be obtained with the application of gibberellic acid. Harvesting of immature fruit can be obtained about 100 days after sowing. In some areas, harvest is year-around. A single plant can produce hundreds of fruit, but 75–100 fruit per plant is common. A planting can yield fruit for several years, but commercial production plants are usually kept for less than 3 years because of attack by nematodes, fungi, viruses and spider mites. In Mexico and Guatemala, starchy tubers are harvested from 2-year-old plants. In China, shoots about 20 cm long are clipped and sold in markets as edible greens.

Sicana odorifera (Casabanana)

Sicana odorifera, one of three species in its genus (Lira-Saade, 1995), derives its specific epithet from its fragrant flowers and fruit. Casabanana, which is probably native to Peru or Brazil, is cultivated in Central and South America and the Caribbean.

This species is a monoecious perennial with stems measuring 15 m long or more. The yellow flowers are solitary, large and relatively tough, persisting for several days after anthesis, during which time they turn pinkish-orange. The cylindrical fruit are yellow, red, dark green or dark purple and measure 30–60 cm long. The juicy, yellow flesh smells sweet, but is tough.

Casabanana is grown as a fruit-vegetable as well as for ornamental purposes. When mature, the fragrant fruit are used to scent linens and clothes, are hung up as room deodorizers and are said to repel insects. They are also made into preserves, but the immature fruit are better for cooking as well as eating raw. Various plant parts have medicinal applications.

Casabanana is propagated from seeds or cuttings and the vines are trained on sturdy trellises or other supports. The crop requires high temperature for fruit ripening. Ripe fruit can be stored for several months.

Telfairia spp. (Fluted Pumpkin and Oyster Nut)

Telfairia occidentalis is called fluted pumpkin because of the ten prominent ribs on the fruit. This West African native is cultivated in Nigeria, Ghana, Sierra Leone and other tropical African countries. Vines grow up to 15 m long, with foliage that is tinted purple. Fluted pumpkin is a dioecious perennial, although Akoroda *et al.* (1990) reported monoecious sex expression. The white, fringed flowers occur in long racemes at the ends of branches on male plants and are solitary on female plants. The pale green ellipsoid fruit, which have a white waxy bloom when mature, grow to 1 m long. The red ovate seeds are up to 5 cm in diameter.

This species is grown primarily for the nutritious stem tips and leaves, which are boiled and eaten. Male flowers are sometimes added to the cooked greens. The large seeds, which have an almond flavour, are prepared for consumption in various ways.

Fluted pumpkin is grown from seeds or vine cuttings. Because the seeds produce multiple shoots upon germination, the large cotyledons of a young seedling can be divided and sectioned so that each division (up to four) is complete with a shoot, some roots and part of a cotyledon.

The vines, which are trained on to supports or grown along the ground, respond well to fertilizer (see 'Fertilizer' in Chapter 6 for more details) and irrigation. Although fluted pumpkin plants are best raised during the rainy season (April–October) in West Africa; irrigation can be used to extend the growing period into the dry season. When the plants are grown primarily for edible greens, young fruit should be removed as they appear. Female plants are hardier and more productive as a leafy vegetable than male plants. Frequent harvesting of young shoots on mature plants encourages branching. Fifteen harvests can be made during a 5-month harvest period, which begins about 3 months after sowing. When seeds for consumption are desired, the fruit should be picked just before maturity.

Oyster nut (*T. pedata*) is another dioecious perennial native to and cultivated in Africa. Vines grow to 35 m long. Leaves have five to seven leaflets. Tendrils are bifid, with one branch longer than the other. Flowers are purplish-pink with fringed margins. The green, ridged fruit have a constricted neck and a distinct rim at the base. They can reach 1 m long and weigh up to 13 kg. The fruit splits at maturity while still on the vine, releasing as many as 200 large brown seeds. The seeds faintly resemble an oyster, hence the common name of the species.

Although the vine tips and leaves are cooked and eaten, this species is grown primarily for its seeds. The seed coats are bitter and must be removed before eating the seeds, which are consumed raw, roasted, pickled, in soup, or for sweets. Edible oil can be extracted from the seeds as well.

Trichosanthes spp. (Snake Gourd and Pointed Gourd)

Snake gourd (*Trichosanthes cucumerina* L. var. *anguina* (L.) K. Pradheep, D.R. Pani & K.C. Bhatt), also known as serpent gourd, is an annual creeping cucurbit native to India. It is commercially cultivated in Asian countries such as India, Bangladesh, Sri Lanka, China, Thailand and Japan. *T. cucumerina* var. *cucumerina* is the wild form of this species. Although similar in name, snake gourd is not the same as snake melon (Armenian cucumber), which is *Cucumis melo*.

Snake gourd is monoecious; male racemes and solitary female flowers appear on the same plant on the same node. The flowers have a fimbriated corolla.

The white, fringed flowers are fragrant and open at night. Snake gourd gets its name from its very long (up to 150 cm) and slender fruit. Two fruit colours are common: light green with white stripes, and dark green with pale green stripes (Fig. 5.16). Open-pollinated cultivars derived from local landraces are most popular with Indian farmers. Some of these include 'Konkan Sweta', 'Kaumudi', 'Harithasree', 'Manusree', 'TA-2', 'TA-19', 'Baby (TA-23)', 'PKM-1', 'MDU-1', 'CO-1', 'CO-2', 'CO-4', 'PLR (SG) 1' and 'PLR (SG) 2'. Hybrid cultivars available in India include 'Snaky White Short' (Ashoka Seeds), 'MHSN 1' (Mahyco) and 'BSS 694' (Beejo Sheetal).

Pointed gourd (*T. dioica*), also known as 'parwal', differs from snake gourd by being dioecious and having smaller fruit (ca 12 cm long) (Fig. 5.17). It is a perennial vine, with simple or bifid tendrils and solitary flowers. It is cultivated

Fig. 5.16. Snake gourd: fruit variation in Indian cultivars derived from local landraces. (Photo credit: Kerala Agricultural University, India.)

Fig. 5.17. Pointed gourd: fruit skin colour variation in Indian cultivars. (Photo credit: Kamal Kumar Yadav.)

primarily in India on approximately 10,000 ha, where a single planting is maintained to produce two annual crops. Propagation is usually by stem or root cuttings. Fresh vines with eight to ten nodes per cutting are selected for planting. A female–male ratio of 9:1 is optimum. Important cultivars in India include 'Swarna Alaukik', 'Swarna Rekha', 'Arka Neelachal Kirti', 'Shankolia', 'Konkan Harita', 'Rajendra Parwal 1', 'Rajendra Parwal 2', 'Faizabad Parwal 1', 'Faizabad Parwal 3', 'Faizabad Parwal 4', 'CHES Hybrid 1' (resistant to melon fruit fly) and 'CHES Hybrid 2'.

6

CULTURAL REQUIREMENTS

INTRODUCTION

Specialized cultivation techniques for cucurbit crops are almost as old as the crops themselves. From a Chinese agricultural encyclopaedia of the 6th century AD comes this description of planting melons, as translated by Shih (1962):

> How to plant melons: Wash the seed with water. Mix with table salt (treatment with table salt protects the vine from mildew). Slice off the topmost layer of dry soil with shovel. (If not sliced off first, no matter how big the hole is, there is always some dry soil in it, and the seeds will not sprout.) Then make a pit as big as a bowl. Place 4 melon seeds and 3 soya beans in the sunny side of the pit. After several leaves have expanded on the melon vine, pinch off the bean seedlings. (Melon seedlings are too weak to break through the ground. Therefore, advantage should be taken of the lifting force of sprouting beans. When melon vines have grown somewhat, the bean-stalks must be got rid of so that the vine may not be shadowed to disadvantage. – Bean-stalks when pinched will give out sap to moisten the soil. Never pull, or the ground would be loosened and dry.)

Even today, the diversity in cucurbit crops is reflected by a diversity of regional cultivation practices. Although many of these practices are converging towards efficacious commercial production techniques, differences among the crops, even at the level of cultivars, requires that individual attention be given to the cultivation of specific cucurbits. Consequently, cultural practices relating to the minor crops are primarily described with each crop in Chapter 5. In this chapter, we discuss research on the cultural requirements of cucurbits, most of which has been conducted on the major crops, and the commercial growing practices associated with these crops.

EFFECTIVE SEED GERMINATION

Seed treatment can help prevent damping off as well as other diseases. Seed companies often coat cucurbit seeds with fungicide and adjuvants to enhance germination and vigour. Hot water (ca 50°C) seed treatment of cucumbers controls some seedborne diseases, as well as promoting germination in freshly harvested seeds. Other cucurbit seed is too large to use hot water treatment effectively, as the outer layers may be killed before heat penetrates to the interior.

Time to germination is usually longer and germination percentages are lower for the minor cucurbit crops (except chayote) than for the major crops. When germinating cucurbit seeds indoors, they should be kept lightly moist, but not too wet, and incubated at low light levels or in darkness. Generally, seed germination is inhibited at temperatures below 15°C and is rapid at 25–30°C. Cultivars within a species can vary greatly, however, in response to temperature. Bushy birdsnest-type melons have relatively large seeds that germinate more quickly and in much higher percentages at 15°C than seeds of other cultivars (Fig. 6.1). Germination of the birdsnest lines can be enhanced further by several treatments, including pre-soaking the seeds for 24 h in aerated distilled water or in aqueous solutions of 3% KNO_3, 6-benzylamino purine (0.1 mmol l^{-1}), or gibberellic acid (1 mmol l^{-1}) (Nerson *et al.*, 1982).

Cucumber seeds imbibed for 6 days at 2.5°C were still able to germinate normally, although emerging radicles were irreversibly damaged after 96 h at this temperature (Jennings and Saltveit, 1994). At 25°C, imbibition appeared

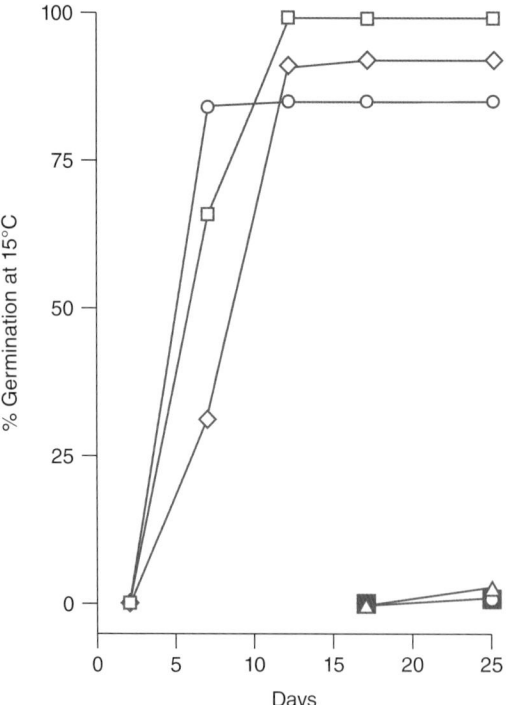

Fig. 6.1. Percentage germination at 15°C of three lines of birdsnest-type melons (top group of symbols) and three other melon cultivars ('Noy Yizre'el', 'Top Mark' and 'U.C. Perlita Bush') (bottom group). Two additional cultivars ('Perlita' and 'U.C. Top Mark Bush') tested did not germinate within 25 days. (Data from Nerson *et al.*, 1982.)

complete between 12 h and 16 h. Longer imbibition periods at this temperature produced seedlings that were more sensitive to subsequent chilling at 2.5°C (Fig. 6.2). The mean germination rate (i.e. the inverse of mean time to germinate) for seeds of 'Dasher II' and 'Poinsett 76' at temperatures ranging from 5°C to 30°C was similar for the two cultivars, but significantly higher than for seeds of 'Poinsett 76' that were 3 years older (Fig. 6.3).

In another experiment (Benzioni *et al.*, 1993), seeds of African horned melon germinated best at 20–35°C and were unaffected by salinities of NaCl up to 50 mmol l⁻¹. Germination was completely inhibited at 8°C. Although there was eventually 90% germination at 12°C, maximum germination was not reached until the 24th day of treatment versus the third day when seeds were kept at 25°C.

Freshly harvested cucurbit seeds may be dormant. Dormancy-breaking treatments, when needed, include scarifying, cracking or removing the seed coat. Clipping the radicle end of the seed can promote germination in bottle gourd and some species of squash. Pre-treatment with gibberellins may aid germination in cucumber, melon, bitter gourd and other cucurbits. Pre-drying is helpful for bottle gourd, cucumber and the squash *Cucurbita pepo*, whereas pre-soaking is advised for angled luffa and watermelon. For recalcitrant seeds of *Cucumis*, pre-chilling at 5°C for several weeks is suggested. In cucumber, secondary dormancy may be induced when trying to germinate seeds under unfavourable conditions.

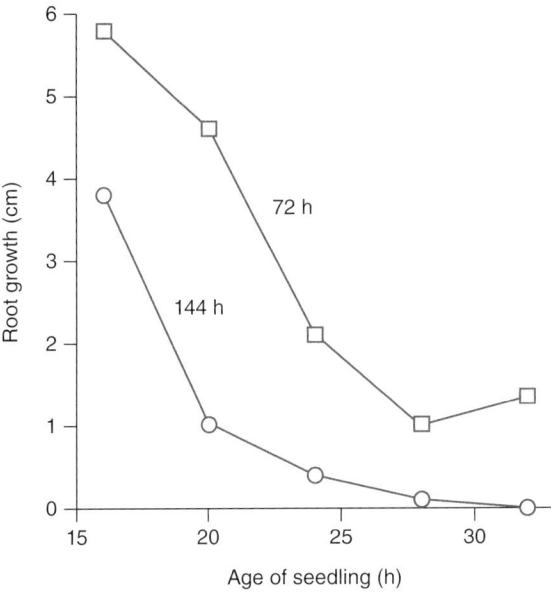

Fig. 6.2. Effect of seedling age (duration of germination of 'Poinsett 76' cucumber seeds at 25°C) on root growth after germinated seeds were chilled at 2.5°C for 72 or 144 h. (Redrawn from Jennings and Saltveit, 1994, with permission from the American Society for Horticultural Science, 600 Cameron Street, Alexandria, VA 22314, USA.)

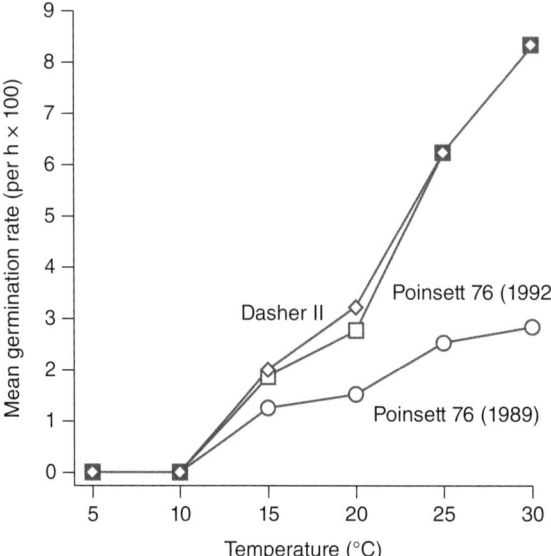

Fig. 6.3. Effect of temperature on mean germination rate of 'Dasher II' and 'Poinsett 76' cucumber seeds from 1992 and 'Poinsett 76' seeds from 1989. (Redrawn from Jennings and Saltveit, 1994, with permission from the American Society for Horticultural Science, 600 Cameron Street, Alexandria, VA 22314, USA.)

LAND PREPARATION

A field crop is usually ploughed following harvest and given time to decay (over the winter in temperate areas). Squash and cucumber have been grown successfully using no-till planting, in which the previous crop and weeds are not ploughed under, but instead are killed with a herbicide and the cucurbit seeds are planted through the stubble (Pierce, 1987).

Tractors and other heavy equipment should not be used for soil preparation or other operations if the field is wet. Rototillers of light weight are commonly used for soil preparation and cultivation in home gardens as well as on small commercial farms in Asian countries.

A winter cover crop can be planted in the autumn to add organic matter to the soil and help prevent erosion. One month before planting the new cucurbit crop, the field should be deeply ploughed and allowed to settle. A fine-grained flat surface is achieved by mechanical disking and harrowing. The field should be relatively free of stones. Levelling may be needed for furrow irrigation.

Raised beds (10–50 cm high) improve drainage, modify temperature and increase the depth of the rooting zone. They can be laid out with a double mouldboard plough to which a marker is attached. The plough makes furrows 25 cm deep and the bed is created by back-furrowing with a turn-plough. The south side of the bed is disked and harrowed and smoothed with a V-drag.

Before planting, the ploughed field can be treated with herbicide to kill germinating weeds, particularly perennial grasses. After the weeds have died, the soil may be fertilized with organic material or inorganic preparations. Inorganic fertilizers are broadcast before beds are formed, banded into the beds between the seed-row and the irrigation furrow, or added to the irrigation water. Irrigation should be applied, if needed, soon after planting and should soak the top 1–2 m of soil.

PLANTING, THINNING AND SPACING FOR OUTDOOR CROPS

Cucurbits are generally planted in the field after the danger of a late frost has passed. Winter squash, pumpkin, melon, watermelon, bottle gourds and luffa require long growing seasons (Table 6.1). Cucumber, summer squash and bitter gourd grow quickly in warm weather, requiring less than 2 months from seeding to the harvest of edible immature fruit. In California and other areas with a long growing season, it is possible to double crop and have two plantings of mechanically harvested pickling cucumbers in the same field. In the frost-free tropics, two crops per year can be produced for wax gourd, and mature chayote plants will bear fruit year-round.

Monoculture of cucurbits is prevalent today. In the past, Native Americans interplanted squash with maize and bean and this is still practised in parts of Latin America. Intercropping is also practised in Africa, where fluted pumpkin is grown with cassava and yam, and egusi is interplanted with grain crops and yams.

Direct seeding in the field is customary. In less mechanized societies, planting is often by the hill system, whereby three to ten seeds are sown by hand in the same place and later thinned to one to three plants per hill. Although home

Table 6.1. Seed size, fruit maturation time, and recommended spacing for major cucurbit crops.

Crop	Number of seeds per gram	Days from planting to harvest	Days from pollination to market maturity	Spacing (cm)	
				Within row	Between rows
Cucumber, pickling	390	48–58	4–5	20–30	90–180
Cucumber, slicing	390	62–72	15–18	20–30	90–180
Melon	460	85–110	42–46	30	150–210
Pumpkin	4–11	100–120	65–110	90–150	180–240
Squash, summer	5–14	40–50	3–6	60–120	90–150
Squash, winter	5–14	85–110	55–90	90–240	180–240
Watermelon	7–11	75–95	42–45	60–90	180–240

Modified from *Knott's Handbook for Vegetable Growers*, © 1980, O.A. Lorenz and D.N. Maynard. Reprinted by permission of John Wiley & Sons, Inc.

gardeners and others still plant using this method, most large-scale growers in developed countries now sow cucurbits with tractor-mounted precision seeders, which uniformly space the seeds, eliminating the need for thinning.

In short-season areas, the season can be extended by sowing the seeds in greenhouses and growth chambers and transplanting young plants in the field. Transplants should be 3–4 weeks old and they should not be bare rooted. Transplants should be rooted in soilless media or other containers that permit transplanting without disturbing the roots. The use of transplants early in the season, especially when plastic mulch and rowcovers are available, can hasten maturity and help the grower to market the crop early in the season when prices are usually higher.

The rate of seedling emergence is influenced by the depth at which seeds are sown. Growers planting early in the season sometimes sow cucurbit seeds at two different depths. The shallow planting comes up first and can be harvested for the early market unless the seedlings are killed by a late frost, in which case they can be replaced by the deeper planting emerging after the frost.

Summer squash and some *C. pepo* winter squash cultivars generally have a bush habit and are grown at much closer spacings than vining winter squash and pumpkin plants (Table 6.1). Bush cultivars with short internodes and a compact plant habit are also available for melon, cucumber and watermelon. They are grown by home gardeners with limited space, but are seldom raised on commercial farms. Small bush cucumbers (e.g. 'Baby Bush') can be placed as close together as every 5 cm within rows. Suggested within-row and between-row spacings of vining plants of the major cucurbit crops are listed in Table 6.1. Field-grown bottle gourd plants should be sown 1–3 m apart.

Increasing plant densities has been used to enhance cucurbit fruit yields. With melon, high-density planting generally produces a greater number of fruit per area, whereas low-density planting can lead to a larger percentage of unmarketable sunburned fruit. However, as plant density increases, fruit quality (i.e. soluble solids content) and fruit number per plant usually decrease, as a result of reduced leaf area per fruit. Comparing within-row spacings of 30 cm and 60 cm between melon plants, Knavel (1991) found that higher density was significantly correlated with lower total leaf area per plant, lower total plant dry weight, fewer fruit per plant (although fruit size was unaffected) and a higher number of fruit per hectare (Table 6.2). Genotype effects reflected differences among the cultivars tested with respect to canopy architecture: longer internodes and smaller leaves in 'Mainstream' and genotype Main Dwarf provided a greater percentage of plant leaf area exposed to sunlight, whereas less secondary stem branching in genotype Ky-P$_7$ meant fewer potential fruiting sites on the plant.

Although plants of the bushy 'Autumn Pride' (*Cucurbita maxima*) have been grown as close together as 30 × 150 cm (22,000 plants per hectare), plants at this high density produced squashes too small to be marketable and some were misshapen (Loy and Broderick, 1990). In contrast, varying

Table 6.2. Effect of plant spacing on fruit number and leaf area for four genotypes of melon.

Parameter and genotype	Plant spacing (cm)	
	60	30
Number of fruit per plant[a]		
'Mainstream'	5.8	3.6
Main Dwarf	5.0	3.8
Ky-P$_7$	1.8	1.4
'Bush Star'	1.4	1.5
Significance		
Spacing	***	
Genotype	***	
Interaction	*	
Number of fruit ('000) per hectare		
'Mainstream'	35.1	89.6
Main Dwarf	36.4	90.8
Ky-P$_7$	13.8	35.5
'Bush Star'	8.9	30.7
Significance		
Spacing	***	
Genotype	***	
Interaction	ns	
Leaf area ('000 cm^2) per plant[b]		
'Mainstream'	15.5	10.5
Main Dwarf	10.8	7.0
Ky-P$_7$	17.0	11.1
'Bush Star'	11.8	5.4
Significance		
Spacing	**	
Genotype	**	
Interaction	ns	

[a]Means for five harvests.
[b]Calculated at 72 days after seeding.
ns, *, **, *** = Non-significant, or significant at $P \leq 0.05$, 0.01, or 0.001, respectively.
Modified from Knavel (1991), with permission from the American Society for Horticultural Science, 600 Cameron Street, Alexandria, VA 22314, USA.

population density from 50,000 to 850,000 plants per hectare did not affect the percentage of misshapen cucumbers for the pickling cultivars 'Bounty' and 'Premier' (Cantliffe and Phatak, 1975). For these cultivars, significant increases in fruit production per area were achieved when planting density was increased from 250,000 to 500,000 plants per hectare (Table 6.3). As with melon, the number of fruit per plant decreased with increasing population density.

Table 6.3. Effect of plant density on fruit yield for two cultivars of field-grown pickling cucumbers.

Plants per hectare	Spacing (cm)	Number of fruit per plant		Tons of fruit per hectare[a]	
		'Bounty'	'Premier'	'Bounty'	'Premier'
850,000	10 × 10	1.7a	1.3a	32.9b	52.3d
650,000	13 × 13	2.0a	1.8a	30.7b	52.3d
500,000	15 × 15	2.1a	2.0a	32.9b	43.3cd
250,000	19 × 19	2.9ab	2.8b	24.0ab	34.4bc
200,000	23 × 23	2.9ab	3.3b	20.3a	33.2b
150,000	25 × 25	4.7b	5.1c	23.5ab	36.3bc
100,000	31 × 31	5.5b	5.5cd	17.9a	31.9b
50,000	46 × 46	8.5c	6.3d	19.1a	23.0a

[a]Harvest made 49 days after seeding.
a–d: Mean separation within columns by Duncan's multiple range test at the 5% level (means followed by the same letter are not significantly different). Modified from Cantliffe and Phatak (1975), with permission from the American Society for Horticultural Science, 600 Cameron Street, Alexandria, VA 22314, USA.

In another pickling cucumber experiment, this one involving an indeterminate vine cultivar ('Tamor') and a highly branched dwarf determinate cultivar ('Castlepik'), Widders and Price (1989) found a high correlation between leaf lamina dry weight and fruit growth rate (Fig. 6.4), supporting the hypothesis that net photosynthetic capacity limits a plant's fruit production potential. Leaf area and fruit number and weight per plant decreased significantly when planting densities were increased from 44,000 to 194,000 plants per hectare (Fig. 6.5). Within-row spacing effects (29, 14 and 11 cm between plants) were more significant than between-row spacing effects (71 and 36 cm). Unlike the results of Knavel (1991) on melon and Loy and Broderick (1990) on squash, genotypic differences in plant architecture did not significantly affect fruit production per cucumber plant. For parthenocarpic (seedless) pickling cucumbers grown in the open field, the optimum density was between 49,000 and 74,000 plants per hectare.

In high-density situations, limited and insufficient resources (e.g. water and nutrients) are expected to create competition among neighbouring plants, tending to decrease the production of an individual plant. At later stages of maturity, especially during fruiting, canopy overlap and shading create competition for light as well. Given that unit leaf rate (total above-ground plant dry weight produced per day divided by the total lamina dry weight) was unaffected by spacing in their experiments, Widders and Price (1989) proposed that foliar competition did not alter photosynthetic efficiency of the leaves, but instead influenced carbon partitioning among alternative sinks within the plant.

Competition, allelopathy and various environmental and genetic factors influence the success of using higher plant densities to enhance overall crop yield. Reduction in individual plant yield and cost of seed must be weighed

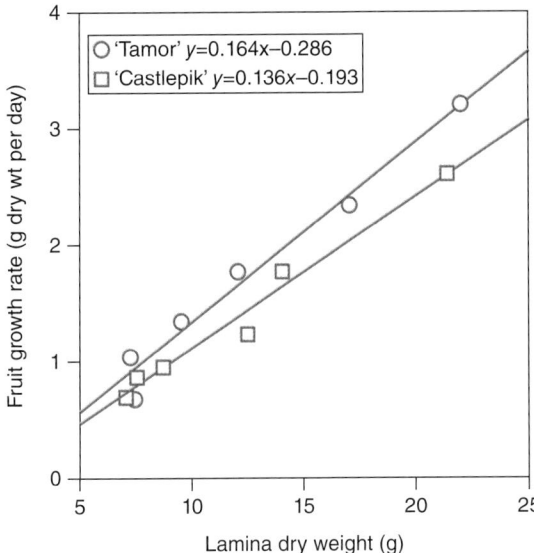

Fig. 6.4. Relationship between total leaf lamina dry weight and net fruit growth rate per plant at 48 days after planting for two cultivars of pickling cucumbers. Correlation coefficients for 'Castlepik' and 'Tamor' are $r^2=0.985$ and $r^2=0.991$, respectively. (Redrawn from Widders and Price, 1989, with permission from the American Society for Horticultural Science, 600 Cameron Street, Alexandria, VA 22314, USA.)

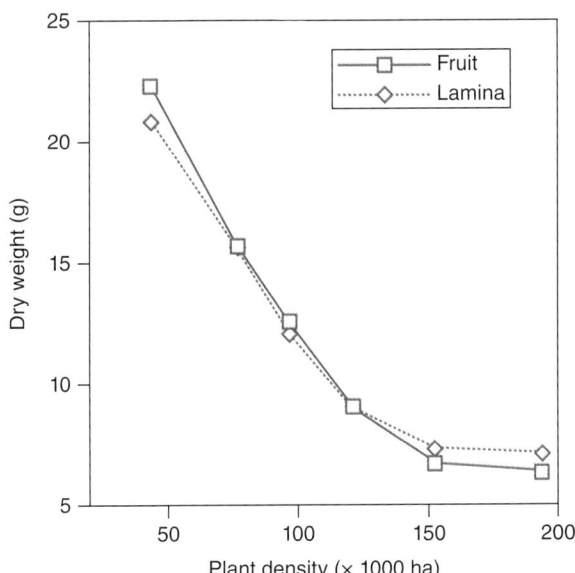

Fig. 6.5. Relationship between population density and the dry weights of fruit and lamina tissues of individual pickling cucumber plants as determined at 48 days after planting. Weight values are means of two cultivars × two replications, and were calculated at six different plant densities. (Modified from Widders and Price, 1989, with permission from the American Society for Horticultural Science, 600 Cameron Street, Alexandria, VA 22314, USA.)

against the increase in plant number per area. For once-over harvested pickling cucumbers, additional concerns include endogenous and exogenous factors that affect the sink strength of early fruit and the number of fruit that can be set simultaneously.

Rows of vertical trellises for luffa production should be at least 2 m apart to accommodate equipment. The results of Davis (1994) on the effects of within-row trellis spacing and pruning on luffa sponge size and quality are presented in Table 6.4. Sponge size, fibre density and sponge appearance scores were greater for plants spaced 91 cm apart than for spacings of 30.5 cm or 61 cm. However, the highest marketable yields were obtained with the 30.5 cm spacing regime.

Chayote plants should be spaced 3–4 m apart on trellises. Bottle gourd, snake gourd, fluted pumpkin and wax gourd plants can be placed every 60–80 cm along the trellis, and bitter gourd plants can be placed as close as 30 cm apart.

MULCHES AND PLANT COVERS

Plastic mulch is used to warm the soil, conserve moisture, reduce leaching of nutrients, suppress weeds and provide protection against soil pathogens and fruit rot. Clear plastic increases soil temperature more than black plastic, but black polyethylene plastic is used more often in order to control weeds. Infrared-transmitting plastic permits infrared rays to penetrate the soil, thereby increasing soil temperature intermediate to that with black or clear plastic mulch and yet still preventing weed growth by the lack of other wavelengths of light.

Aluminium mulch or aluminium bonded to paper reflects heat as well as light and does not warm the soil as well as black or clear plastic. However, aluminium and other brightly coloured mulches have the advantage of repelling aphids and some other insects, thereby reducing losses from viral diseases transmitted by these insects.

Plastic mulch can be applied with a tractor-mounted implement before planting. The plastic covers the row for a width of 1 m or more; at intervals, small openings are made in the plastic to permit insertion of seeds or transplants into the soil. The openings are generally sufficient for water absorption from rain or irrigation, but it is recommended that the plastic be applied only when there is adequate soil moisture. Plastic mulch is often used in combination with rowcovers and drip irrigation.

Polyethylene plastic needs to be removed and disposed of at the end of the season. Biodegradable and photodegradable plastic mulches have been developed to avoid this problem.

In the USA, hot tents (small paper coverings over individual plants), often supplemented with windbreaks of brush, have been used to protect melon

Table 6.4. Effect of within-row spacing and pruning on size and quality of luffa sponges.

Main effect	Mean sponge size Diameter (cm)	Mean sponge size Length (cm)	No. sponges/ha ('000s) Diameter 5.1–7.6	Diameter 7.7–10.2	Length 15.2–30.5	Length 30.6–45.7	Fibre density[a]	Sponge strength[b]	Sponge appearance[c]
Within-row spacing (cm)									
30.5	7.9	39.9	36.2	31.2	10.7	42.6	2.4	2.2	2.6
61	8.3	42.2	24.0	32.4	7.4	31.7	2.6	2.3	2.4
91	8.6	42.7	15.7	33.7	4.1	28.8	2.9	2.6	2.9
LSD$_{0.05}$	0.02	1.5	5.9	ns	3.0	5.8	0.4	ns	0.4
Pruning									
Topping at node 6	8.3	42.7	22.1	30.9	5.5	31.0	3.0	3.0	2.9
Removing four laterals	8.2	40.6	28.6	31.1	8.5	37.2	2.6	2.0	2.6
No pruning	8.3	41.6	25.2	35.3	8.1	34.8	2.3	2.1	2.4
LSD$_{0.05}$	ns	1.5	ns	ns	ns	ns	0.4	0.4	0.4
Significance									
Within-row spacing	**	**	**	ns	**	**	*	ns	*
Pruning	ns	*	ns	ns	**	ns	**	**	*
Interaction	ns	ns	ns	ns	ns	*	ns	ns	ns

[a]Overall fibre density rated on a scale of 1 to 5 (1 = very loose, open weave of fibres through which newspaper print could be read; 5 = very closely woven fibres that allowed little light to penetrate).
[b]Sponge strength was rated on a scale of 1 to 5 (1 = crumbled easily; 2 = did not crumble but could be pulled apart; 3 = took some effort to pull apart; 4 = was very difficult to pull apart; 5 = could not be pulled apart).
[c]Visual appeal was rated on a scale of 1 to 5 (1 = poor with loose, broken, uneven fibres; 3 = looked good with fairly closely woven, even fibres; 5 = excellent with very closely woven, even fibres and a smooth finish).
LSD = least significant difference.
ns, *, **= Non-significant, or significant at $P \leq 0.05$ or 0.01, respectively.
Modified from Davis (1994), with permission from the American Society for Horticultural Science, 600 Cameron Street, Alexandria, VA 22314, USA.

plants early in the season, when it may be cool and windy. Similarly, strips of rye, sorghum or other cereal can be sown in rows at right angles to the prevailing wind to protect cucurbit seedlings.

For additional protection from outdoor elements, especially cold, various enclosures exist that extend the growing season of cucurbits in temperate areas. For example, home gardeners may start their melon plants in cold frames (small framed glass structures) that each accommodate a few plants or flats. The soil or other medium in the frame is warmed by sunlight and the absorbed heat keeps the plants warm during cold spring nights. Cloches (glass covers over plant rows) have been used successfully in the UK for many years to increase local temperature and protect plants from frost.

More recently, unheated plastic tunnels have become popular in Mediterranean countries and elsewhere for commercial production of melon, cucumber, summer squash and other cucurbits. Polyethylene or polypropylene plastic is usually suspended by wire hoops over a row of plants to form these rowcovers. Floating non-woven polyester rowcovers require no support: lightweight fabric rests on the plants, with the edges secured by soil to keep it in place. Tunnels and floating rowcovers increase temperature during the day, protect young plants from the wind and exclude harmful insects and insect-transmitted diseases. However, these coverings provide only slight protection against frost.

The temperature may build up in plastic tunnels so much on warm sunny days that cucurbits are harmed unless ventilation is provided. Some growers attach strips of plastic to wire hoops with clothes pins or other devices so that the plastic can be opened on warm days. Others use plastic with slits for ventilation. To permit insect pollination, the rowcovering should be removed or opened before cucurbits flower unless a parthenocarpic cultivar is grown.

Soltani et al. (1995) evaluated various mulch and rowcover combinations for their effects on growing watermelon. Although light intensity was lower under rowcovers, the three materials used (clear polyethylene, spun-bonded polypropylene polyamide net and spun-bonded polyester fabric) transmitted enough of photosynthetic photon flux (70–80%) that the light saturation point of the plants was probably exceeded. Although carbon dioxide (CO_2) concentration was higher inside the mulch transplanting holes (600 µl l^{-1}), foliage level concentrations were normal, apparently due to rapid airflow inside the tunnels. Consequently, temperature was deemed the most important environmental factor modified by the treatments. Compared with bare ground, soil temperatures were significantly higher under mulches (ca 2–4°C higher at 10 cm soil depth under black mulch) and more so (by an additional 1–3°C) when rowcovers were used with the black mulch. As a result of warmer temperature, plants under rowcovers (especially those made of polyethylene or polyester) grew faster, flowered sooner and yielded more fruit.

GLASSHOUSE CULTURE

The largest protective structures used for commercial cucurbit production are glasshouses, which, broadly speaking, include angled-roof buildings constructed of aluminium and glass, structures made of corrugated fibreglass and rigid or air-inflated houses enclosed with polyethylene. Glasshouses typically possess equipment to control one or more of the following environmental conditions: supplementary lighting (e.g. light-emitting diode (LED) lamps of white or other colours), shading (e.g. inner or outer blinds), heating (e.g. hot water or air), cooling (e.g. evaporative cooling pads), ventilation and humidity (e.g. mechanically controlled ceiling vents) and CO_2 concentration. Irrigation set-ups range from overhead sprinklers to drip fertigation systems to sophisticated hydroponics.

Glasshouse cultivation is suitable for most cucurbits when trellising, or bush cultivars can be used. In China, Japan and Korea, some melon and watermelon cultivars are commonly raised in glasshouses. However, the most suitable cucurbit crop for growing in a controlled environment is cucumber (Fig. 6.6). In fact, cucumber may have been the first crop protected from unfavourable outdoor weather; the plants were grown in forced culture by the ancient Romans. Today, glasshouse cucumber is an important crop in The Netherlands, the UK, China, Japan, Korea and the Middle East. Most of the discussion of glasshouse culture that follows is based on experience with and research on cucumber.

In order to conserve glasshouse space and heating, cucumber is often grown from transplants. Seedlings are cultivated in small containers to about 10 cm high and then transplanted at a wider spacing (ca 40 cm apart within a row) for trellising. Given their climbing nature, these plants are usually raised to maturity in beds, boxes, troughs, pots or bags that sit on the clean glasshouse floor or on low perforated benches. The roots are anchored in sterilized soil or a soilless medium of peat, vermiculite, perlite, straw, sawdust, sand or other substrate, such as rockwool, a relatively inert material composed of coke and limestone made molten at 1600°C, spun into fibres and then woven into slabs.

The use of rockwool culture is popular in The Netherlands and the UK. The slabs are laid on grooved polystyrene boards in which a plastic water pipe is set for root zone warming. For planting, holes are cut on top of the slabs through the polyethylene sleeve. Frequent irrigation is necessary with rockwool and root zone pH should be kept at 5.5–5.7.

In straw bale culture, fertilizer is added to water-soaked bales, which are then subjected to at least 15°C to promote fermentation. After bale temperature peaks near 40°C and starts to fall, cucumber plants are rooted in a soil or peat mixture placed on top of the bales. Straw from wheat or barley crops treated with herbicides should be avoided, as the residues can harm cucumber plants. This is an older system that requires extra labour, but reduces heating costs.

Fig. 6.6. Cucumber plants growing in a glasshouse for production.

With some growing systems (e.g. peat modules and rockwool slabs), nutrients can be supplied with the irrigation water via a fertilizer injector in an 'open' watering system. In a closed hydroponic set-up, the aqueous nutrient solution is recycled, monitored and adjusted as necessary. For example, Japanese producers raise cucumber and melon seedlings in perlite for 4–6 weeks until the young plants can support themselves. At that point, the plants are inserted into the styrofoam lids of solution-filled troughs (Resh, 1987).

The nutrient film technique (NFT) is a hydroponic set-up that strives to improve aeration by constantly circulating a shallow nutrient solution across the cucumber root zone. The solution is pumped from a catchment tank up to the raised ends of sloped troughs made of polyethylene film. The solution flows through the troughs past the roots of the inserted plants and back into the tank. NFT systems can be complex, with solution-heating tubes to maintain a high root temperature (at least 18°C and preferably 25–29°C for cucumber), CO_2 enrichment lines, collecting troughs, and electronic control devices for monitoring the pH (5.5 is good for cucumber) and electrical conductivity of the solution.

Glasshouse temperatures should be no lower than 16°C at night, ranging up to 30°C during the day, depending on the crop and stage of growth. An average daytime temperature of 21–22°C, with 2°C less at night and venting at 27°C, is standard glasshouse practice for winter cucumber crops in northern Europe. After plants are well established, mean temperature can be dropped by 2–3°C. Heating to higher temperatures, which would produce greater yields, is not considered cost effective when outdoor temperature is low (Liebig, 1980). However, higher growing temperatures are possible for summer crops and are recommended for glasshouse-grown melon.

Ventilation is required, especially in summer, to reduce glasshouse temperature and humidity. High humidity, which builds up as a result of transpiration, promotes fungal growth. At the end of a hot day, venting prevents an increase in relative humidity during a subsequent cold night. If glasshouse temperature rises too quickly in the morning, venting is needed to avoid condensation on fruit that are often cooler than the surrounding air. Air movement that encourages transpiration is desired and accomplished with passive ventilation through roof openings or with the aid of fans.

Supplementary lighting and CO_2 enrichment in the glasshouse help to increase cucumber yield and reduce time to harvest. Fluorescent or LED lamps are kept on over cucumber seedlings for 12 h or more per day for the first 3 weeks of growth, producing larger and hardier seedlings. Supplementary lighting can be continued until the flowering stage, at which time adjustments in daily light duration may be needed to accommodate the plant's sensitivity to daylength.

The benefits of supplementary lighting are enhanced further by the concurrent application of supplemental CO_2, since both are necessary resources for photosynthesis. For the pickling cucumber 'Wisconsin SMR 18', the greatest yield in dry weight of young plants was obtained at the highest levels tested: 2150 ppm CO_2 combined with light at 6000 μW cm^{-2} (Hopen and Ries, 1962). As expected, the interaction of light and CO_2 was significant. Plants grown to fruiting developed a greater number of marketable fruit at higher light levels.

The effect of CO_2 enrichment alone was evaluated for glasshouse cucumber crops grown in North Carolina (Peet and Willits, 1987). Extended CO_2 duration (by the use of rock storage instead of venting to control temperature) and greater CO_2 intensity were tested. At short durations (< 6 h per day), concentrations of 5000 μl l^{-1} produced up to a 30% increase over controls (i.e. no CO_2 enrichment) in marketable fruit yields of spring crops. At longer enrichment durations, maximum results were reached at lower concentrations (600–800 μl l^{-1}).

In commercial cucumber production, standard practice is to increase glasshouse CO_2 levels to 1000 μl l^{-1} during those daylight hours when vents are closed, or nearly so; higher levels are not considered cost effective. Carbon dioxide gas can be obtained and injected directly into the glasshouse, or produced by burning low-sulfur kerosene, paraffin, propane or natural gas. Cucumber plants grown on bales of decomposing straw, which emit CO_2 at nearly $1000 \, \mu l \, l^{-1}$, do not need supplemental CO_2 (Hand, 1984).

An advantage of greenhouse cucumbers is their ability to set early parthenocarpic fruit without pollination. A single incompletely dominant allele, plus modifying genes, is the basis for this parthenocarpy. These alleles were derived from European germplasm developed more than a century ago. Breeders selected high-yielding types in their winter glasshouses, without realizing that genetic parthenocarpy was the reason for the high yields produced by plants grown in the absence of pollinating insects. Also, parthenocarpic fruit are less inhibiting to additional fruit formation on a plant than are fruit with seeds.

Cucumber yields can be considerably higher in glasshouses than in the field. Contributing factors are parthenocarpy, closer spacing, more favourable climate and fewer insects and diseases. Thus, glasshouse cucumbers are harvested more times than field cucumbers. Three to five glasshouse crops can be produced a year, or one long-duration crop can be grown in the glasshouse from February through October by using the renewal umbrella system of trellising.

GRAFTING

Cucurbit grafting became popular in Japan and Korea in the 1920s. Most watermelon, melon and glasshouse cucumber crops are grafted in these countries (Table 6.5). Grafting is also practised in other parts of Asia (e.g. China) (Figs 6.7, 6.8) and Europe where land use is very intensive, but was not popular in the USA until 2000.

Table 6.5. Annual statistics for the total area cultivated in cucurbits and the area cultivated with grafted seedlings for Japan and Korea in the early 1990s.

	Japan		Korea	
Cucurbit crop	Total cultivation area ('000 ha)	Area of grafted plants ('000 ha)	Total cultivation area ('000 ha)	Area of grafted plants ('000 ha)
Watermelon				
Field	24.9	23.8	28.3	26.9
Glasshouse	3.2	3.2	7.4	7.4
Cucumber				
Field	14.8	4.6	3.8	0.4
Glasshouse	7.0	6.0	4.7	3.3
Melon				
Field	5.5	3.7	3.9	3.3
Glasshouse	3.4	2.5	5.2	5.0

Modified from Lee (1994), with permission from the American Society for Horticultural Science, 600 Cameron Street, Alexandria, VA 22314, USA.

Fig. 6.7. Cucumber seedling grafted to Malabar gourd rootstock in Chinese glasshouse. Note grafting clip holding the hypocotyls together below the large cotyledons of the rootstock.

Fig. 6.8. Grafted watermelon seedlings.

Grafting a crop to the rootstock of another cucurbit cultivar or species can provide resistance to soilborne diseases, increase low temperature tolerance and enhance water and nutrient uptake, which promotes growth and extends harvest time (Table 6.6). Fruit of grafted watermelon can be significantly larger than those of self-rooted plants. Rootstocks must be carefully chosen, however, because they may have additional undesirable effects on the grafted scion. For example, a cultivar-specific trait such as fruit colour in cucumber can be modified by the biochemical products of the rootstock. The interspecific squash cultivar 'Tetsukabuto' (*Cucurbita maxima* × *Cucurbita moschata*) is used as a rootstock for watermelon and other cucurbits in Asia.

Table 6.6. Cucurbit rootstocks, grafting methods and purposes of grafting for cucurbit crops.

Cucurbit crop	Popular rootstocks	Grafting methods[a]	Purposes[b]
Watermelon	Bottle gourd	1	1,2
	Interspecific hybrids[c]	1,2	1,2,3
	Wax gourd	1,3	1,2
	Cucurbita pepo	2,3	1,2,3
	Cucurbita moschata	1,2	1,2,3
	Sicyos angulatus	2	5
Cucumber	Malabar gourd	2	1,2,3
	Interspecific hybrids[c]	1,2	1,2,3
	(*Cucurbita maxima* × *C.moschata*) F_1	2	1,2,4
	Cucumber	2	1,2
	Sicyos angulatus	2	2,5
Melon	Interspecific hybrids[c]	2	1,2,3
	Cucurbita moschata	2	1,2,3
	Melon	2,3	1,3,4

[a]Grafting methods: 1 = hole insertion; 2 = tongue approach; 3 = cleft grafting.
[b]Purposes of grafting: 1 = fusarium wilt control; 2 = growth promotion; 3 = low temperature resistance; 4 = growth period extension; 5 = nematode resistance.
[c]Many interspecific hybrids are commonly obtained by fertilized ovule culture *in vitro*.
Modified from Lee (1994), with permission from the American Society for Horticultural Science, 600 Cameron Street, Alexandria, VA 22314, USA.

For current information on rootstocks, consult the Vegetable Grafting database (Anonymous, 2018).

Cucurbits are usually grafted before expansion of the first leaves in rootstock and scion seedlings. The following grafting descriptions are from Lee (1994). 'Hole insertion' grafting is frequently used for watermelon scions because of their small size as seedlings. The apical bud is clipped out of the rootstock; the watermelon seedling is severed just above the roots and then inserted into the hole between the cotyledons of the rootstock. The 'tongue approach' is often used with cucumber seedlings because of their hearty hypocotyls. A downward and upward cut is made in the hypocotyl of the rootstock and scion, respectively. The 'tongue' of the scion is then inserted into the tongue of the rootstock, and a grafting clip secures the area. After about a week, the hypocotyl of the scion is cut below the graft union. 'Cleft grafting' is employed for older seedlings with leaves. A cleft-shaped segment is removed from the side of the top of the severed rootstock stem. The scion is clipped above the root area, placed in the cleft of the rootstock and secured with a special grafting clip. All grafted seedlings need to be monitored for 7–10 days after grafting. Commercial grafted seedling production continues to expand, due to reduced costs from automated methods of grafting.

TEMPERATURE

Cucurbits are sensitive to low root-zone temperature, which increases suscepti-bility to diseases. At temperatures below 20°C, water uptake may be restricted and the plant injured or killed by drought, even though soil moisture is ample. When sowing cucurbits on ridges or raised beds, planting on the southern slope of the bed will give the seedlings the advantage of somewhat warmer soil temperatures. For winter glasshouse crops, water absorption may be restricted by the low temperature of the irrigation water. To prevent this problem, some growers heat the cold irrigation water or blend it with warm water. Increasing the temperature of the medium around the roots is also practised. This can be accomplished with a circulating hot-water system constructed of poly-ethylene pipes buried in the medium. Water at 35–40°C is used to maintain root medium temperature at 20–23°C.

Kramer (1942) investigated the relationship between soil temperature and transpiration in watermelon plants, finding that only 20% and 50% as much water was absorbed at 10°C and 15°C, respectively, as at 25°C. Decreased absorption at low temperatures appears to be the result of several factors, including decreased root growth, lower respiration rates, increased viscosity of colder water and decreased permeability of root cell membranes. Root per-meability in watermelon decreases when temperatures fall below 22°C, with water uptake decreasing most rapidly between 18°C and 16°C.

Cucurbit plants are subject to chilling injury when exposed to temperat-ures less than 10°C, causing yellowing of foliage and short misshapen fruit. Cucumber fruit become pitted with small sunken spots and rot prematurely. Chilled winter squashes and pumpkins do not store well, but will rot prema-turely also. Plants can be 'hardened', principally by adequate exposure to low but non-injurious temperatures, to somewhat reduce the temperature at which chilling injury occurs. For example, several days exposure to 12°C can harden cucumber plants to subsequent temperatures below 10°C.

Cucurbits require warm temperatures (20–30°C) for best growth and the annual species are often the first vegetables to be killed by a frost. The an-nual stems of perennial species (e.g. pointed gourd) may be killed by frost also, but new vines sprout from surviving underground stem tissue when warm weather returns. Malabar gourd and stuffing cucumber are more cold tol-erant than most cultivated cucurbits, being able to grow at temperatures only slightly above freezing. Among the squashes, *Cucurbita maxima* and *C. pepo* are more tolerant of low temperature than are *C. argyrosperma* and *C. moschata*, which thrive at high temperature. In fact, fruit of some cultivars of *C. maxima* and *C. pepo* taste sweeter when grown in temperate areas with cool nights. In the tropics, these species can be cultivated at elevations of up to 2000 m. Cucumber prefers warm conditions, but can be grown at cooler temperatures than melon or watermelon. Melons are of best quality when raised at 25–30°C. Stable day/night temperatures promote rapid growth in watermelon.

In various experiments (e.g. Krug and Liebig, 1980), raising the 24 h mean temperature from 15°C to 30°C caused cucumber stem and leaf growth rates to accelerate, time to first harvest to shorten, and an increase in total fruit yield (Fig. 6.9). Although the amplitude around the mean generally had little effect, young plants (< 34 days old) responded better to higher day temperatures among regimes with the same 24 h mean (Slack and Hand, 1983). However, large quick fluctuations should be avoided. A sharp drop in temperature can cause cucumber fruit to become thin around the middle and a short period of low temperature can produce scars on developing fruit. In cucumber, there is a nuclear gene (*Ch*) for high tolerance to chilling and a cytoplasmic gene for slight tolerance to chilling. Most cucumbers are susceptible to chilling when treated at the second true leaf stage at 4°C for 7 h in the light. Damage will be less if chilling occurs in the dark, or if the chilling duration is shorter, the chilling temperature is higher, the air speed is lower, or the plants are younger.

Grimstad and Frimanslund (1993) found that stem elongation in young cucumber plants was less when night temperature exceeded day temperature. As long as fruit yield is not affected, maintaining a relatively short height is desired for greater ease in transporting and transplanting young plants in the glasshouse.

Optimum temperatures contribute to crop yield primarily when they occur during the early part of the growing season. The developmental response to

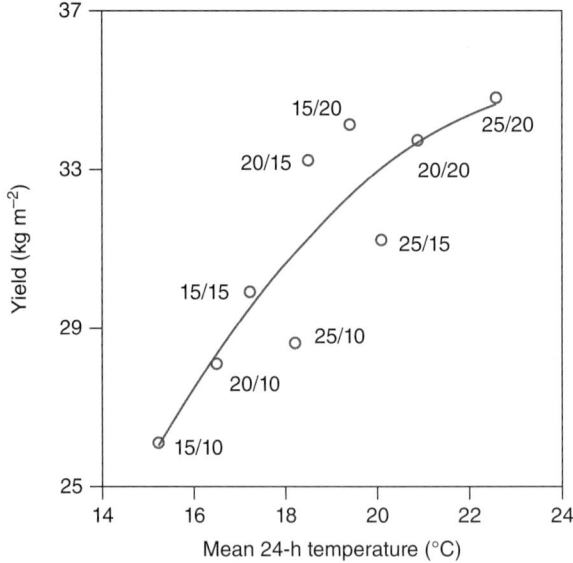

Fig. 6.9. Effect of mean 24 h temperature from various day/night temperature regimes on glasshouse cucumber yield after 20 weeks of harvesting. The numbers by each data point indicate the nominal day/night temperature setting for that treatment. (Data from Slack and Hand, 1983.)

temperature decreases as the crop matures, with other factors, especially fruit load, affecting fruit growth and yield on older plants. Cucumbers germinate and grow fastest at 32°C and will go from planting to first harvest in 39 days if grown at that temperature.

DAYLENGTH AND LIGHT INTENSITY

Growing cucurbits in extreme latitudes is complicated by the more variable photoperiods as well as the cool temperatures in these areas. Long days favour male over female flower production in many cucumber, wax gourd and squash cultivars. In other species (e.g. Malabar gourd and chayote), long days inhibit flower production altogether. Genetic breeding has created various daylength-insensitive cultivars within the major cucurbit crops, and many of the minor cucurbit crops are naturally day-neutral.

Temperate-region growers wanting to produce fruit from short-day plants like chayote during the summer need to control the photoperiod of young plants. From about the third to the fourth month of growth, vines should be shaded in the morning and evening with a dark cloth over a sturdy frame to give the plants a daylength of 8 h. Vines can be allowed to grow under natural daylengths after flowering has begun (Aung *et al.*, 1990).

High light intensity is needed for optimum yield of cucurbits. Although chayote will flower under glasshouse conditions in the winter, fruit set is poor because of the low light intensities. Furthermore, vigorous vine growth under these conditions can cause readsorption of the contents of developing fruit by the vine (Aung *et al.*, 1990). Supplementary illumination in the glasshouse is beneficial in areas and seasons with much cloudy weather. In the field, aluminium or other reflective mulch on the soil reflects additional light to the leaves of cucurbit crops.

Sunburn injury of immature fruit happens when cucurbit vines are moved aside during harvesting, exposing previously shielded tender fruit to high light intensity. A white or tan necrosis develops on the part of the fruit suddenly exposed to sun.

SOIL

Cucurbits benefit from well-drained soils with a high organic content. Sandy or sandy loam soil is preferred for early production, but cucurbits can be grown on a variety of soil types. Fields with sufficient air drainage may escape frost at the beginning or end of the season, resulting in a longer growing period. A heavy soil with good moisture-retaining properties may be preferred in a relatively dry non-irrigated situation. However, very heavy soils that are wet for extended periods or those compacted by large machinery should be avoided, since adequate soil aeration is needed.

Cucurbits will not tolerate 'wet feet'; if their roots are submerged for too long, the plants will be wilted and stunted and may not fully recover, even when drainage is later improved. In experiments on summer squash, waterlogging reduced the number and length of lateral roots, but promoted the development of adventitious roots at the stem base near the soil surface (Huang *et al.*, 1995). Salinity in the water exacerbated the negative impacts of waterlogging on root, shoot, leaf and fruit growth. When 14 days of waterlogging were followed by 7 days of drained conditions, the squash plants were able to recover from most of the adverse root effects.

Most cucurbits should not be planted on highly acidic soils. Soil pH of 5.6–6.8 is recommended for cucumber, squash, luffa, chayote and wax gourd. Watermelon tolerates pH as low as 5.0, but grows best at pH 6.0–7.0. Soil pH 6.0–6.7 is recommended for melon plants; they grow poorly and may become chlorotic when grown in very acidic soils. If liming is required, it should be applied long before planting cucurbits to give soil acidity time to adjust.

FERTILIZER

Concerns in industrial countries over water pollution by fertilizer and manure have led to a renewed interest in crop rotation with soil-improving, nitrogen-fixing legumes. Singogo *et al.* (1996) found that fruit yields of the melon 'Magnum 45', grown after a single winter cover crop of alfalfa or Austrian winter pea, were similar to yields produced with application of inorganic nitrogen at 100 or 135 kg ha^{-1}. Furthermore, the addition of feedlot beef manure to the soil before ploughing under the cover crop in spring produced only slight increases in the yield of the summer melon crop.

It has long been the practice in Egypt and India to apply manure in trenches before filling the trench with soil and planting watermelon. In most areas today, a complete fertilizer with nitrogen, phosphorus and potassium (NPK) is generally ploughed down before planting cucurbits. The type and amount of fertilizer varies in different growing areas. For cucumber, Pierce (1987) recommended 22–34 t ha^{-1} of animal manure or green manure equivalent. He reported that application amounts of inorganic fertilizer for cucumber ranged from 56 kg ha^{-1} each of N, phosphoric anhydride (P_2O_5) and potash (K_2O) in northeastern states of the USA, to 168 kg nitrate, 134 kg phosphate and 202 kg potash per hectare in Florida. All or part of this fertilizer is ploughed down and the rest banded at planting.

Scientists tested fertilizer formulas for growing fluted pumpkin in acidic sandy soil in Nigeria (Obiagwu and Odiaka, 1995). They found that phosphorus was the most important component and that P alone at 22 kg ha^{-1} gave higher fresh-matter yields than the same amount of P along with N or K. In their experiments, band application or broadcasting was significantly more effective than ring application or point placement methods.

Side-dressing with supplementary N may increase yield, especially for cucurbits having a long growing season, but it is not recommended for cucumber plants that are to be harvested once by machine. Furthermore, ammonium nutrition is detrimental to cucumber plant growth at soil pH < 5.0 or > 7.5. Hollow heart, in which an empty cavity develops in the fruit flesh, and excessive vine growth of watermelon can be caused by large amounts of N fertilizer. In situations where supplementary N is desirable, applications should be made before flowering, until the vines become too large to permit a tractor between the rows.

Supplementary N and other nutrients can be provided through the irrigation water (a process called fertigation) if plastic mulch or thick vine growth prevent side-dressing close to the plants. Fertigation allows the use of smaller plant beds than those used with band application. When applied via irrigation, a smaller amount of fertilizer can produce the same or greater yields as produced with a larger amount of dry granular fertilizer. Swiader *et al.* (1994) found that sprinkler-applied 112N:112K split into five fertigations during the growing season (and supplemented with a pre-plant application of dry 28N:56K) yielded large numbers of marketable *C. moschata* pumpkins without compromising early maturity. Increasing potassium to 224 kg ha^{-1} produced favourable results as well. However, lower amounts of nitrogen (e.g. 56N:112K) or higher amounts of either nutrient (e.g. 168N:224K) gave smaller fruit yields as a result of later flowering and, in the case of limited nitrogen, poor growth and lower overall flower production (Table 6.7).

Table 6.7. Effect of N:K fertilizer regime on pumpkin (*Cucurbita moschata*) female flowering, percentage fruit set, and plant dry weight at 72 days after seeding.

N:K fertilizer[a] (kg ha^{-1})	Number of days to first female flower	Number of female flowers per plant per day[b]	Female flower fruit set (%)	Plant dry weight (g plant^{-1})
1987 Trial				
56:112	60.0a[c]	0.07b	40.3b	100c
112:56	51.8b	0.12a	53.5a	241b
112:112	51.0b	0.13a	51.9a	271b
196:280	53.0b	0.09ab	57.8a	357a
1988 Trial				
112:112	49.0b	0.14a	53.3a	316b
112:224	50.0b	0.13a	52.8a	339b
168:224	56.0a	0.13a	58.4a	420a
196:280	52.8ab	0.09b	60.2a	481a

[a]All fertilizer treatments were applied by fertigation except for 196N:280K, which was a dry-blend fertilizer applied before planting.
[b]For the period from initial anthesis to 72 days after seeding.
[c]For each year, mean separation within columns by LSD at *P*= 0.05.
Modified from Swiader *et al.* (1994), with permission from the American Society for Horticultural Science, 600 Cameron Street, Alexandria, VA 22314, USA.

Scientists in Florida conducted trials on N requirements for slicing cucumbers grown with polyethylene mulch and drip irrigation (Hochmuth and Hochmuth, 1991). Testing treatments of $0-222$ kg ha^{-1}, they found that total fruit yields levelled off with the application of 153 kg ha^{-1} in 1988 (under conditions of leaching rainfall) and 64 kg ha^{-1} in 1989, when conditions were considerably drier.

In glasshouse trials with melon, Sharples and Foster (1958) found that plant dry weight was highest at about 0.26 g of added ammonium nitrate per kilogram of local soil. At that level of N, additions of mono-ammonium phosphorus and potassium sulfate had little effect on growth. The scientists went on to test the effects and interactions of these nutrients with calcium (Ca) and magnesium (Mg). In the presence of high available N (0.66 g added per kilogram of soil), high Ca content in the soil created Mg deficiency (< 1%) in the leaves; calcium ions also competed with potassium ions to restrict K concentration in leaves (Fig. 6.10). When available N was decreased, low N and high Ca interacted significantly to reduce plant growth and Mg levels in leaves.

Adams *et al.* (1992) tested the interactions of N, K and Mg fed to glasshouse cucumber plants grown in sphagnum peat. As expected, increasing amounts of N or K caused increasing Mg deficiency and corresponding losses in fruit yield.

Mineral element deficiencies, including sulfur (S), iron (Fe), manganese (Mn), boron (B), copper (Cu), zinc (Zn), silicon (Si) and molybdenum (Mo), can cause chlorosis, stunting, poor fruit shape and other problems. For example, insufficient Ca promotes the development of blossom end rot in cucurbit fruit under moisture stress. The first symptom of Mg deficiency is the yellowing of

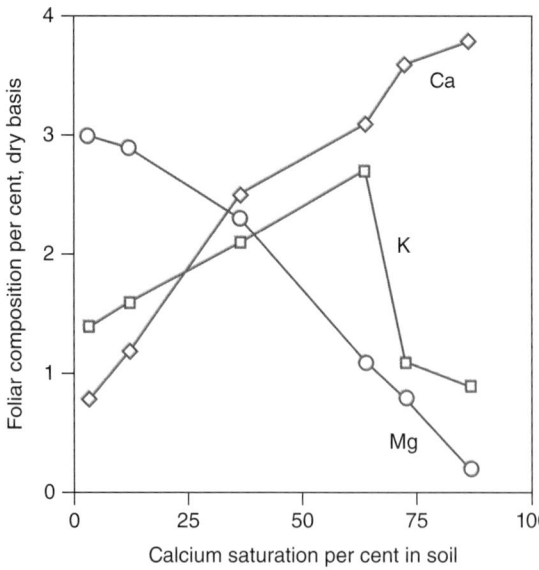

Fig. 6.10. Effect of varying the calcium saturation percentage of the soil on Ca, K and Mg composition in leaves of melon line CBR-1. (Data from Sharples and Foster, 1958.)

the margins of older leaves. Inadequate boron leads to an increase in auxin activity in squash roots, inhibiting their growth. Silicon deficiency in cucumbers reduces vegetative growth and fruit production and increases susceptibility to some mildew diseases. Symptoms of various element deficiencies in cucumber were described and illustrated by Rordan van Eysinga and Smilde (1969) and Zitter *et al.* (1996).

Foliar sprays are an effective means of delivering most minerals. For example, a low-volume spray with 10% magnesium sulfate is suggested for cucumber plants showing signs of Mg depletion. Alternatively, Mg can be added to the nutrient solution (ca 50 ppm) or soil (ca 50 kg ha^{-1}). Melon plants grown on sandy or acidic soil respond to Mg provided in dolomitic lime.

Excessive concentrations of Mg, K, Ca, S and Mn stunt plant growth (Fig. 6.11). Manganese toxicity is sometimes a problem in freshly steamed glasshouse soil; leaf symptoms in melon include pinhole lesions and water-soaked margins on the lower surface.

When nutrients are applied with irrigation water, pH and electrical conductivity (EC) of the solution affect the availability and toxicity of the nutrients to plant roots. For example, Fe, Mn, cobalt (Co) and Zn become less available as alkalinity increases above pH 6.5, whereas Mo, S and K become less available as acidity increases below pH 6.0. Mg and Ca are most available to plant uptake at pH 7.0–8.5.

For adequate cucumber nutrition, the total concentration of elements in solution should be around 1300 ppm, giving an EC reading of about 2.0 dS m^{-1}. Although higher EC (2.5–3.0) can produce stronger plants and better fruit colour in low light situations in glasshouses, values above 4.0 may suppress growth.

Fig. 6.11. Cultivar differences in response to manganese toxicity: (left) 'Wisconsin SMR 18'; (right) 'Ashley'.

Genetic differences in melon for salt tolerance have been reported, but no cucurbit grows well in highly saline media. High total salts content (especially sodium chloride (NaCl)) restricts uptake of certain elements (e.g. Fe), causes mineral toxicities (e.g. Na, Cl, B) and increases tolerance to other elements, such as Zn. Secondary effects are inhibited root development, reduced growth, dark green foliage with a marginal necrosis and lower yield. When summer squash plants were watered with saline solution (NaCl at 100 mmol m^{-3}) for 21 days, the number of fruit per plant was 45% lower than for plants grown under non-saline conditions (Huang *et al.*, 1995). In melon, increasing salinity in the EC range of 1.2–6.0 caused fruit yields to decrease as a result of decreasing fruit size (Mendlinger, 1994). However, increasing salinity also produced higher percentages of soluble solids in fruit. Similarly, cucumber plants grown in saline conditions of about 3000 ppm salts may yield 25% less than normal, but produce sweeter-tasting fruit.

IRRIGATION

Cucurbits, with their long vines and large leaves, have high rates of water loss via transpiration. Consequently, they often benefit from supplementary irrigation, especially when grown on light sandy soils. Field-grown cucumber plants may need water every 5–7 days and are usually irrigated, even in humid areas. Heavy soils that have been soaked just before planting to a depth of 2 m may contain sufficient moisture for average crops of melon, squash and watermelon without the need for additional irrigation. Vining cultivars of squash and pumpkin may wilt on hot dry afternoons if not irrigated, but recover turgidity the next morning. Watermelon and wax gourd have deep roots and can survive relatively dry conditions. However, watermelon and other cucurbit fruit can develop a blossom end rot during periods of moisture stress. Early evidence of serious water stress in cucurbits includes the loss of colour in the lobes of older leaves and the tips of developing fruit, and irreversible wilting.

Water stress affects all major physiological processes, from photosynthesis to carbohydrate metabolism. It also causes the hydrolysis of protein. In pumpkin, wilted plants exhibit large increases in free proline, which appears to function as a storage compound during times of stress.

Genetic variation for drought tolerance exists in various cucurbit species. The *C. pepo* zucchini squash 'Dark Star' was selected through a farmer participatory breeding project to develop a summer squash cultivar that could be used for dry farming in California (Desclaux *et al.*, 2012).

Cucurbits are usually grown on raised beds if furrow irrigated and on level soil with other irrigation practices. Raised beds dry out faster than flat rows and require more frequent watering. In the USA, commercial growers of field crops often use mobile overhead sprinklers. The supporting equipment may be self-propelled and the sprinklers slowly rotating to ensure total field coverage. In flat fields with a high water table, irrigation water that is added to the soil through ditches or pipes becomes available to plant roots via capillary action in the soil. This seepage system reduces fertilizer leaching. Drip or trickle irrigation has become popular for cucurbits in Israel and other areas where water conservation is important. Water slowly emanates from emitters on plastic pipes directly to the root zone of cucurbits, with little water loss by run-off or evaporation. Compared with furrow, seepage or sprinkler irrigation, drip irrigation conserves water and helps prevent accumulation of soluble salts in the soil. It may also result in fewer foliar diseases than overhead irrigation practices, which leave free water on the leaves.

To prevent fruit rot, irrigation should be avoided as fruit near maturity. Sudden irrigation or flooding following a period of drought can cause fruit to crack, making them more vulnerable to pests and pathogens. Melons of the Momordica Group are very prone to cracking, which usually happens near fruit maturity. Excessive soil moisture also causes plant wilting and stunting.

Glasshouse-grown cucumbers should be thoroughly watered for the first few days after transplanting, then irrigated less for about 2 weeks to promote healthy root formation. Thereafter, a regular irrigation schedule can be adopted, for example 15 min of drip irrigation every 2 h from dawn to dusk in the summer, depending on temperature. Irrigation should be performed during the day, when transpiration rates are high. Water or solution application should not be so great or of such long duration that drainage is impeded, restricting root access to oxygen. Irrigation scheduling can be managed with the use of a tensiometer to monitor water potential of the soil or other medium.

Pier and Doerge (1995) evaluated the combined effects of soil water tension and N fertilizer on trickle-irrigated watermelon plants grown in Arizona fields. Predicted maximum marketable fruit yield was 102 t ha^{-1} at 7.2 kPa tension (measured at 0.3 m depth) and 336 kg N ha^{-1} (supplied as urea via fertigation in five applications). However, this combination of water and fertilizer inputs created unaccounted for, or 'lost', N estimated at about 60 kg ha^{-1}, an amount that could cause unacceptable contamination of groundwater. When unaccounted-for N loss was limited to < 15 kg ha^{-1} and the costs of water and fertilizer were taken into account, 95% of the maximum yield and net return values occurred with an average soil water tension of 7–17 kPa and N applications totalling 60–315 kg ha^{-1}.

SUPPORT, TRAINING AND PRUNING

Field-grown cucumber plants are often allowed to grow freely on the ground, whereas glasshouse cucumbers are usually trained vertically or at an angle to conserve space. Cucumber plants in the field are sometimes grown on trellises for the same reason. Also, long-fruited cultivars must be trained on a trellis since fruit over 230 mm in length will curve when grown flat on the ground. Chayote, bottle gourd, luffa, bitter gourd, small-fruited wax gourd cultivars and other minor cultigens with large climbing vines (e.g. species of *Trichosanthes*) are commonly grown on outdoor trellises, particularly in Asian countries.

A trellising system can be made of long strips of fencing or be as simple as three leaning poles joined together at their apices. In a system composed of vertical support posts and a single horizontal top wire or pole, growers suspend strings from the horizontal support to help train the plants to the top of the trellis. Cucurbits with large heavy fruit (e.g. wax gourd or large-fruited bottle gourd cultivars) should be supported by very sturdy trellises.

In the glasshouse, a V-cordon system trains cucumber plants to grow upwards at an angle in order to make best use of light. Two support wires positioned up to 1 m apart are anchored about 2 m above a row of plants. Plastic strings lead some branches up to one support wire and other branches to the other wire. Plastic clips secured to the string are positioned every 30 cm or so along the branches under leaf petioles for vine support.

The renewal umbrella system trains glasshouse cucumbers vertically until they reach and are secured to the support wire. Then, the growing tip of the main stem is cut to encourage lateral branching. All lateral branches are removed except for two at the top of the plant, which are trained over the support wire and then down along opposite sides of the main stem. Cucumbers are allowed to develop along the top third of the main stem and on the two laterals. As these fruit mature, two more laterals are allowed to grow from the top of the plant; all other new laterals are removed. When fruit are ready for harvest, the branches on which they occur are cut and new fruit are allowed to develop on the next set of hanging laterals. This process can be repeated over a 10-month period, with single plants producing more than 100 fruit (Resh, 1987).

Pruning vegetative parts and flowers can affect female flower earliness and the quality of fruit on a plant. Terminal shoots on bottle gourd plants should be cut off when they are about 3 m long to promote lateral branching and female flower production. Topping the main stem increases sponge length and quality in fruit of smooth luffa (see Table 6.4), whereas pruning laterals promotes flowering. Melon plants grown in cloches are often pruned to keep the plants within the restricted growing area and to remove excess fruit. However, leaf pruning of melon plants after fruit set has begun can impair yield and fruit quality.

If too many fruit are allowed to develop per plant, quality may be adversely affected. For glasshouse melon cultivars, leaves, lateral branches and fruit are sometimes pruned so that only one fruit per lateral and a total of four fruit per plant are maintained. Some watermelon growers remove all but two fruit per plant in order to increase fruit size and uniformity. The superfluous fruit should be removed when still quite small. More fruit per plant can be left on the vines of small-fruited watermelon cultivars. Squash and pumpkin growers striving to produce fruit of enormous size permit only one fruit to develop on a plant.

WEED CONTROL

Weed control is essential for good yield of cucurbits and it makes harvesting easier. Cultivation, hoeing and other mechanical means of tillage are used. Deep cultivation should be avoided, since cucurbits have many shallow roots. Once the vines are well established and cover the ground, their abundant foliage canopy intercepts light and inhibits growth of late-germinating weeds.

Genetic differences in allelopathy have been found and proposed as a means of weed control for cucurbits. Roots of some lines of cucumber reportedly release chemicals that inhibit germination of weed seed in the surrounding soil. This characteristic has not been bred into any cultivar, however, and is not used commercially to aid weed control.

Black plastic and infrared-transmitting mulches are often used to prevent weeds from growing.

Cucurbits generally are very sensitive to some herbicides, but a few cultivars may be more tolerant. For example, they can be injured by atrazine residues in the field from a previous maize crop, or by very small residues of clopyralid (3,6-dichloropicolinic acid) that linger after application to pastures or food crops, or from compost made from hay treated with clopyralid before it was harvested. Although ethalfluralin is registered for cucurbits in the USA, misuse of this and other dinitroaniline herbicides is a common cause of injury to cucurbits (Zitter *et al.*, 1996). Wet cool conditions may increase the chance of injury from these and some other herbicides (e.g. chloramben). A herbicide injurious to cucurbits, such as glyphosate, can be applied to the soil between strips of plastic mulch if directed away from the cucurbit plants.

There are other herbicides on the market that can be used safely in cucurbit fields. For example, many cucurbits are tolerant of the pigment inhibitor clomazone. The sulfonylurea herbicide halosulfuron is widely used on many cucurbits. Herbicides are applied either pre-planting and incorporated, or after planting but before emergence. Sufficient soil moisture is required for various of these herbicides to be effective; and in the case of those applied after planting, irrigation or rainfall is needed to 'activate' the herbicide. Better and more selective herbicides are needed for cucurbits.

The 'stale seed bed' technique has been employed to prevent weeds from becoming a problem in cucurbit fields. Several weeks before the cucurbits are to be sown, the field is ploughed and prepared for planting. This encourages weed seeds to germinate and then the weeds are killed with a non-residual herbicide (e.g. glyphosate) before the cucurbit seeds are planted. When glyphosate is used for control, it should be applied to weeds 3 days before cucurbit seeding.

Soil fumigation is effective for weed control, and it also controls soilborne diseases and nematodes. It is seldom employed for cucurbits in the field because of the cost and environmental concerns, but has been used for cucumber plants grown in glasshouse ground beds.

For disease, insect and weed control chemicals, refer to the local manual, such as the North Carolina Agricultural Chemicals Manual (https://content. ces.ncsu.edu/north-carolina-agricultural-chemicals-manual).

FRUIT AND SEED PRODUCTION

CONTROLLING SEX EXPRESSION

Although some summer squash cultivars can set fruit without pollination, many cannot. Poor fruit set in the latter group is a common complaint early in the flowering season, when there may be insufficient male flowers for reliable pollination. Breeders have selected summer squash lines that produce many pistillate and few staminate flowers. This can contribute to early fruit set and has reduced the cost of producing hybrid seed in cases where emasculation is accomplished by manually removing male flowers from the female parent. However, it has created problems when predominantly female cultivars are planted early in the season, because low temperatures promote female sex expression even more, to such a degree that male flowers for pollination may be lacking. This problem can be overcome by interplanting a highly pistillate cultivar with a predominantly male cultivar (e.g. 'White Bush Scallop'), but this is seldom done. Instead, most growers simply wait until the cultivar planted produces both staminate and pistillate blossoms.

The use of gynoecy to produce hybrid seed without the need for emasculation has become extremely important for cucumber, though not for melon. In fact, gynoecious cultivars represent one of the most important developments of cucumber breeding in modern times. Since the gynoecious inbred parent of a hybrid cultivar generally has only female flowers, the open-pollinated seeds it produces are F_1 hybrids. Thus, costly hand pollination and emasculation are not necessary.

Most gynoecious hybrid cucumber cultivars are heterozygous for F, the allele for female sex expression. Under some environmental conditions, these cultivars produce male as well as female flowers. Cultivars stable in gynoecious sex expression can be bred by selecting for the right combination of modifying genes with Ff, or for the homozygote (FF). Homozygous gynoecious hybrid seed has been created by hand pollinating two gynoecious inbreds in isolation, usually in a glasshouse, after treating the male parent with a growth regulator to induce male flower formation. However, this method of producing

homozygous seed is expensive. A more efficient method is to use a hermaphroditic line (which has the genetic background *FF mm*) as the male parent for the
hybrid. The parental lines are grown in adjoining rows and cross-pollination is
accomplished by bees instead of by hand. Hybrid seed is harvested only from
the gynoecious inbred (*FF MM*) used as the female parent. The hybrid is homozygous gynoecious because both parents were homozygous for the gynoecious
allele *F*. Its being heterozygous for the andromonoecious allele *m* is of no con-
sequence, since this allele is recessive and has no effect on the hybrid.

If a gynoecious cucumber hybrid is not parthenocarpic, its seed is usually blended with 10–15% of seed from a monoecious hybrid (or inbred) to
ensure sufficient male flowers for pollination. Originally, 'Sumter' was used
as a pollenizer for single-harvest, gynoecious field cultivars for pickling; its
first staminate flowers open 30–35 days after planting. Walters and Wehner
(1994) found several sources of breeding material in which anthesis begins
within 26–30 days after planting. Pollenizers for pickling and fresh market
types are usually monoecious hybrids, produced using two monoecious
inbred parental lines.

In some cucurbits (e.g. bottle gourd), removing male flower buds will increase vegetative growth, including the production of lateral branches, where
a greater percentage of female flowers occur. However, this method of promoting femaleness is not effective in cucumber (Delesalle and Mooreside,
1995).

Many growth regulators have been reported to influence sex expression.
Ethylene is a naturally occurring plant hormone that plays an important role
in this and many other physiological processes in cucurbits. The effects of light
intensity, photoperiod and other environmental factors on sex expression may
be mediated through their influence on endogenous ethylene.

Ethylene is generally associated with femaleness. For example, gynoecious
cucumber plants generate more ethylene than monoecious plants. Exogenous
application of ethylene, or compounds that release ethylene, can induce a bisexual floral bud to develop into a pistillate flower in cucumber, melon and
squash. However, ethylene has been shown to promote maleness in watermelon (Rudich, 1990), and inhibitors of ethylene action or synthesis such as
silver nitrate, silver thiosulfate, or 2-aminoethoxyvinylglycine promote hermaphroditic flowering in monoecious watermelon (Christopher, 1982).

The growth regulator most widely used to stimulate production of pistillate flowers is ethephon. The active ingredient of this growth regulator,
2-chloroethylphosphonic acid, is an unstable compound. When its pH increases, such as when it is applied to cucurbits, this compound breaks down to
release ethylene. Monoecious cucumber and squash plants treated in the seedling stage with ethephon produce only pistillate flowers for an extended time
and have reduced internode lengths (Robinson *et al.*, 1969, 1970). However,
ethephon treatment of smooth luffa seedlings caused no change in sex expression in some experiments (Wehner and Ellington, 1995).

Ethephon is widely used to facilitate production of F_1 hybrid summer squash seed. This seed was formerly produced by hand pollination or by manually removing staminate flower buds from the inbred line used as the maternal parent, with bees bringing pollen from a male parent grown in an adjoining row. Now, ethephon eliminates the need for laborious removal of staminate flower buds to prevent selfing or sibbing of the maternal parent. Several rows of the ethephon-treated female parent are usually grown for each row of the male parent in seed production fields. After fruit set is completed, plants of the male parent are eliminated from the field so that their seeds will not become mixed with those of the F_1 hybrid.

Squash lines differ in response to ethephon. Plants with a high degree of female sex expression, i.e. having a high ratio of pistillate to staminate flowers when untreated, are generally easy to convert to gynoecy. Several applications of the chemical are usually required during the flowering season to prevent the female parent from producing male flowers. Shannon and Robinson (1979) determined that two applications of 400–600 µl l^{-1} significantly inhibited staminate flower development in summer squash without reducing seed yield or quality.

Ethephon has also been applied to the female parent of hybrid cucumber cultivars that are monoecious, but there is no need for it with gynoecious hybrid cultivars unless the female parent produces some staminate flowers.

In order to maintain a gynoecious cucumber line by selfing or sibbing, it can be induced to develop male flowers with the application of another important endogenous regulator of sex expression, gibberellic acid (GA). GA has the opposite effect of ethephon: it typically stimulates development of staminate blossoms, although it has been shown to promote femaleness in bitter gourd. GA_{4+7} is much more effective than GA_3 for stimulating the production of staminate flowers in gynoecious cucumber plants. Also used for this purpose are silver compounds, such as silver nitrate or silver thiosulfate. Silver compounds, aminoethoxyvinyl glycine (AVG) and CO_2 promote maleness by inhibiting the activity of ethylene.

Male-sterile mutants have been found for cucumber, melon, watermelon and squash. In each case, sterility is inherited as a single recessive allele. Since a cytoplasmic factor is not involved, lines that breed true for this character are not possible. This is a limiting factor in the use of male sterility to produce seed. Since the female parent will segregate for male sterility, male-fertile plants must be rogued in seed production fields. All male-fertile plants should be removed from the female parental line before flowering, but this can be difficult since they cannot be recognized until the male flower buds form a few days before anthesis. Glabrous foliage is associated with a male-sterile allele (*gms*) of watermelon, making it possible to classify plants for this allele in the seedling stage. Male sterility has been used to a limited extent in the production of F_1 seed of the squash *Cucurbita maxima*. Although not extensively used, male sterility is potentially advantageous for producing melon seed, since it is tedious and

expensive to remove the anthers from hermaphroditic flowers of the andromonoecious maternal parent of a hybrid. Male sterility can be induced by high
concentrations (> 2.2 mmol l^{-1}) of maleic hydrazide.

Other compounds that can be applied to suppress maleness or promote
femaleness are auxins, 6-benzylaminopurine, flurenol, chlorflurenol, abscisic
acid and boron. In melon, flurenol or chlorflurenol is applied to the foliage of
young plants or the seeds are soaked in the solution for 4–5 days (Roy and
Saran, 1990). Various growth regulators (e.g. 6-benylaminopurine) can also
be introduced through the roots of seedlings.

POLLINATION

Flowers of cucumber, melon, squash and watermelon open in the early
morning and pollination is generally completed by noon. In most of the minor
cucurbits such as *Luffa* spp., wax gourd and bitter gourd, anthesis starts early
in the morning and anther dehiscence precedes flower opening and stigma
receptivity remains maximum at anthesis. Anthesis starts in the evening in
white-flowered species such as bottle gourd and snake gourd. Anther dehiscence begins 2–3 h before flower opening and stigma receptivity is maximum
at anthesis.

Most species of the Cucurbitaceae are pollinated by bees or other insects,
which are attracted to pollen and nectar. Different species of *Peponapis* and
Xenoglossa (gourd bees) preferentially pollinate different species of squash.
Cucumber beetles can pollinate bottle gourd and other cucurbits. Pollination
by hummingbirds and butterflies has been reported for *Gurania* and *Psiguria*,
and moths and bats are believed to pollinate cucurbits with white night-
opening flowers.

Some growers place hives of honey bees in their cucurbit fields when the
plants flower in order to improve fruit set. This gives higher quality fruit or
better seed production, depending on crop objectives. Two or three large bee
colonies per hectare, positioned within 100 m of the edge of the field, are recommended for cucumber. For melon, the bee colonies should be placed in
the field, as opposed to on the borders, after flowering has begun, or the bees
might migrate to other fields. Winter squash and pumpkins need a minimum
of 500 pollen grains per flower for fruit set with yields maximized at around
2500 grains. Exact pollen requirements will vary by fruit size and potential
number of seeds, with larger fruit possessing more seeds and greater quantities of pollen. Depending on the pollinator, these amounts of pollen will be
achieved in 5 (bumble bees) to 10 (honey bees) visits (Pfister *et al.*, 2017).
Recommendations for winter squash production fields in the Pacific Northwest
are for a honey bee hive every 0.4–0.8 ha. In fields less than 120 ha, hives
can be placed at field margins and do not need to be placed within the field
(Gavilánez-Slone, 2000). Care should be taken when applying pesticides, to

avoid killing pollinators. Insecticides with short residual activity are best and should be applied in the afternoon, since pollinating insects are usually in cucurbit fields in the morning.

Bees can bring cucurbit pollen from great distances and cross-pollination with a bitter ornamental gourd (*Cucurbita pepo*) is possible even in isolated seed production fields of that species of squash. An alert breeder will recognize the crosses in the next generation, since hybrids between summer squash and an ornamental gourd will generally have shorter, rounder fruit with warts and different rind colour than is typical for the squash cultivar being grown. Even though the offending hybrid plants are rogued and not allowed to produce seed, bitterness can persist in succeeding generations if bees distributed pollen from the gourd–squash hybrid plants before the plants were removed from the field. The heirloom 'Delicata' winter squash was almost lost because of outcrossing to an ornamental gourd. Re-selection in remnant seed resulted in the release of 'Zeppelin', which preserves the exceptional quality of this type.

As part of their quality control programme, it is recommended that breeders self-pollinate a number of individual plants of open-pollinated squash cultivars or the inbred parents of F_1 hybrid cultivars. After the selfs are grown out and tested for bitterness, remnant seeds of all selfs proved not to carry the allele for bitter fruit can be combined as elite seed stock for future increases of the cultivar.

Cucurbits are self-compatible. Reports of self-incompatibility in squash have been proved to be spurious. Most monoecious cucurbits are naturally self- as well as cross-pollinating. The bisexual flowers of andromonoecious melon and cucumber plants seldom set fruit if not mechanically pollinated. Although they are self-compatible and will set fruit if self-pollinated, they normally require the assistance of a pollinator because pollen is shed towards the outside of the flower, away from the centrally located stigma. When pollinated by insects, hermaphrodite flowers generally have a higher rate of successful self-pollination than do pistillate flowers.

Inadequate pollination can cause poor fruit shape in cucumber. Although pollen normally takes a few hours to reach ovules, if the rate of fruit elongation exceeds the rate of pollen tube growth, then the more distant ovules in long-fruited cucumbers are never fertilized. Also, if there are insufficient pollen grains on the stigma, generally only the ovules closest to the stigma are fertilized. In either case, the blossom end of the fruit is stimulated by fertilization to enlarge more than parts of the cucumber without seeds, resulting in a misshapen fruit.

'Seedless' cucumber cultivars will produce seeds if pollinated. Therefore, seed production for parthenocarpic cucumbers is similar to other types. Seeds produced per fruit may be low if the ovary is very long, due to the distance that pollen tubes must travel to fertilize distant ovules. For fruit production, glasshouse vents can be screened to exclude insects that might pollinate and cause poor fruit shape in seedless cucumbers.

Parthenocarpy (fruit set without pollination) occurs in cucumber. Parthenocarpic cucumbers have been used mainly in the greenhouse as well as high tunnels for production of Dutch greenhouse type and greenhouse Beit Alpha (Persian cucumber) type cucumbers. The fruit are seedless and no pollenizer (monoecious cultivar) or pollinator (honey bee hives) are needed for production.

More recently, parthenocarpic (seedless) cucumber has been used in open-field production of pickling cucumber in Europe and North America. Fresh-market (slicing) cucumber type is also becoming available in parthenocarpic cultivars. It is useful if growers plant the parthenocarpic cultivars in areas away from standard (seeded) types. There will be fewer pollinated (seeded) fruit if bees and foreign pollen are away from the parthenocarpic cucumbers. Usually, 2 km of distance between fields producing the two types of cucumbers is sufficient.

Parthenocarpy can sometimes be induced by rubbing the stigmas with pollen of another species or induced chemically. In cucumber, chlorflurenol (2-chloro-flurenol-9-carbonic acid) is effective for promoting parthenocarpy (Robinson et al., 1971). Gibberellins applied to the stigma can cause seedless parthenocarpic fruit development in chayote (Aung et al., 1990). A modified cytokinin, 1-(2-chloro-4-pyridyl)-3-phenylurea, which promotes fruit set in pollinated flowers of melon and watermelon, also produces parthenocarpy in unpollinated flowers of watermelon (Hayata et al., 1995).

HARVESTING

Minor cucurbit crops, which are grown primarily in non-industrialized countries, are harvested by hand. Two to three months after planting, edible immature fruit of bitter gourd, chayote, wax gourd, angled luffa and bottle gourd can be picked weekly throughout the fruiting season. In temperate regions, smooth luffas grown for sponges, bottle gourds grown for containers and wax gourds grown for storage generally should remain on the vine until the end of the growing season. In tropical areas with year-round growing, smooth luffas are mature when the rind turns brown and the seeds rattle inside the lightweight dry fruit. Bottle gourds can take 4–6 months to mature on the vine, depending on the cultivar; the fruit turns from green to light brown or yellow when mature. Those harvested too soon may have thin rinds and not store well.

Multiple harvests are made by hand for melons and are sometimes performed at night if the days are hot. Melons harvested when immature have low sugar content and poor flavour. However, Conomon Group melons grown for pickling should be picked when immature, once they are of full size.

Maturity of cultivars of the Cantalupensis Group can be gauged by the appearance of the 'slip', (the abscission zone on the peduncle at its attachment to the fruit). A melon is said to be 'full slip' when the abscission layer forms a ring entirely circumscribing the fruit attachment, causing the ripe fruit to detach naturally, or slip, from the vine. Fruit harvested before half slip, when the ring is no more than half completed, may be inferior in quality. They can be recognized by the jagged scar at the stem end of the fruit where it was detached before completion of the abscission layer. For 'PMR 45', half slip occurs 35–40 days after anthesis.

Inodorus Group melons do not form an abscission layer. Their maturity is judged, with some difficulty, by the size, shape, feel and aroma of the fruit. For example, a 'Casaba' melon is deemed ready for harvest when there is a slight yielding at the blossom end when thumb pressure is applied. Although 'Honey Dew' melons can be harvested as soon as 35 days after anthesis, ethylene treatment is usually needed to ripen the fruit sufficiently for market. When left on the vine, the fruit naturally become mature around 50 days after anthesis, which is too late for commercial harvesting since over-ripening occurs within a few days (Pratt *et al.*, 1977).

Watermelon fruit do not separate naturally from the vine and it may be difficult for the inexperienced to judge maturity. Size and colour of the fruit change as maturity approaches, with the fruit becoming larger and fuller and the ground spot, where the fruit rested on the ground, changing from white to light yellow. The tendril closest to the fruit may wither and turn brown when the fruit ripens, but this is not always a reliable indicator. The time-honoured ritual of thumping the fruit is done to determine if the sound is sharp, ringing or metallic, indicating immaturity, or muffled and resonate when the fruit is ripe. Watermelons are of highest quality when harvest is delayed until the fruit has 10% or higher soluble solids content.

Watermelons should not be harvested in the early morning, when they are more turgid and apt to crack. A sharp knife is used to cut the fruit from the vine, leaving about 3 cm of peduncle attached to the fruit to deter stem end rot. The peduncle may be treated with a paste or wax to further limit microbial access.

Winter squashes and pumpkins grown for fresh market and storage should be hand harvested before freezing weather damages the fruit and reduces storage life. A light frost will not hurt mature fruit and may facilitate harvesting by killing the vine cover and exposing the fruit, but prolonged exposure to temperatures below 10°C causes chilling injury. For winter squash destined for storage, freezing temperatures should be avoided since chilling injury will increase storage rots. Likewise, if fruit can be harvested before a major precipitation event, longevity in storage will be improved. In tropical countries, mature squashes are picked when the rind loses its sheen and the tendril nearest the fruit dies. As with watermelons, 2–3 cm of peduncle is often left on the fruit when packing into small cartons or crates. For direct harvest into large pallet bins, peduncle length should be reduced, or the peduncle removed, to avoid damage to other fruit.

Mechanical harvest systems have been developed for winter squash grown for processing and culinary seed. This system has been used with *C. maxima* 'Golden Delicous', *Cucurbita moschata* 'Dickinson Field' and *C. pepo* hull-less seed pumpkins. These possess large spherical to hubbard-shaped fruit. These shapes are preferred because the fruit roll easily and are readily shifted into windrows using tractor-mounted blades. For processing, the windrowed fruit is picked up with conveyer belts and loaded into bins for transport to the processing plant. Where 'Golden Delicious' or hull-less seeded varieties are grown solely for culinary seed, the fruit is cracked, pulp and seeds are collected and the skin and pericarp are returned to the field using a mobile wet seed combine.

Cucumber, summer squash, bitter gourd and other cucurbits grown for their immature fruit are more productive if the fruit are harvested frequently (e.g. every other day) and not allowed to become too large. Large fruit act as a sink for nutrients and inhibit development of additional fruit on the plant. Summer squashes should be picked soon after the blossom falls off; they are tenderer and slightly sweet at this stage. In China, the blossom may be left on slicing cucumbers to indicate freshness of the young fruit. In the USA, pickling cucumbers are sold on the basis of size, the smallest fruit having the greatest value. The proportion of fruit in small size grades is increased if harvesting is performed at frequent intervals.

Although slicing cucumbers are harvested by hand, mechanical harvesters have been developed for pickling cultivars. Some harvesters are designed to make multiple harvests, but the most popular harvesters make only a single harvest in a field. It is important for a once-over mechanical harvest that the cultivar and cultural conditions be such that the maximum number of fruit are ready for harvest at the same time. Gynoecious cultivars with uniform germination are recommended. Planting density should be at least 100,000 plants per hectare. H.C. Price and D.W. Kretchman (unpublished, 1990) recommended that the plants be grown on well-drained uniform soil with less than 1% slope and free of large stones. Long fields are efficient, since they minimize the time spent turning the harvester at the end of the row. Fields with a persistent herbicide residue or heavily infested with weeds should be avoided. Rotation is recommended to avoid planting in areas where cucumber or pepper crops were infected with *Phytophthora* the previous year, or in soils with residues from the previous corn crop. To increase yield, parthenocarpic fruit set can be induced by the growth regulator chlorflurenol, where governmental regulations permit its use.

For once-over harvesting, prediction of harvest date is important and has typically been based on the number of days from planting to harvest of the previous year's crops. Perry and Wehner (1990) tested a heat unit model labelled 'reduced ceiling' for predicting optimum harvest date for pickling and slicing cultivars of cucumber. This model sums, over days from planting to harvest, the difference between the daily maximum (except if the maximum is greater than 32°C, in which case it is replaced by 32°C minus the difference between the maximum and 32°C) and a base temperature of 15.5°C. They found that their heat unit summation value was a better predictor of harvest date than the standard method of counting days for picklers, but not for slicers.

Harvest aids (e.g. tractor-drawn conveyer belts) have been used to improve efficiency in collecting and processing cucumbers, melons, watermelons and other cucurbits grown in large fields. The harvest aid closely follows the pickers so that the workers can easily place the fruit on the conveyer belt, which carries the produce to the packing area.

POSTHARVEST HANDLING AND STORAGE

Fruit

Malabar gourd and wax gourd fruit have the longest shelf-lives among cucurbits grown for consumption. They can be stored for a year or more in a cool, dry environment. Although cucurbit fruit generally keep longer at lower temperatures, they should not be frozen or have prolonged exposure to chilling temperatures (< 10°C, exact temperature depending on the cultivar). A change in respiration that leads to the production of harmful end products (e.g. ethylene) at low temperature appears to contribute to chilling injury. In cucumber, pitted rinds and collapsed flesh are evidence of damage. African horned melon develop opaque rind spots when stored at 4°C. Although fruit of this species will keep for 3 months at 24°C, shelf-life was shortened at lower temperatures in experiments. At 20°C, there was 30% spoilage among stored fruit by day 37 (Fig. 7.1); at storage temperatures below 12°C, all ripe fruit spoiled within 55 days (Benzioni *et al.*, 1993).

Cucurbit fruit harvested immature (e.g. summer squash, cucumber, bitter gourd, angled luffa, chayote) should be handled carefully to avoid damaging their thin skins. Rind wounds give fruit an unmarketable appearance, allow entry of pathogens and reduce shelf-life. Unprocessed, immature fruit usually survive storage for only a few weeks. Bitter gourds keep for about 2 weeks at 12–13°C and 85–90% relative humidity (RH). Summer squashes can be kept safely at 7–10°C and 85–95% RH, but chilling injury may occur at lower temperatures. Cultivars of *C. pepo* summer squash with allele *B* (see 'Genes' in Chapter 3) in their genetic background are more susceptible to chilling injury and shrivelling during storage than similar cultivars lacking this allele. Shrivel symptoms may appear when there is as little as 6% weight loss in fruit of some straightneck squash cultivars (Sherman *et al.*, 1987).

Winter squashes and pumpkins should be fully mature before being harvested for storage (Fig. 7.2). They can be kept in bulk bins, either temporarily before processing or for long-term storage. Slatted bins allow better air circulation and fewer losses to disease than solid-wall bins. Care should be taken not to overfill storage containers, as compression damage to fruit is possible. For

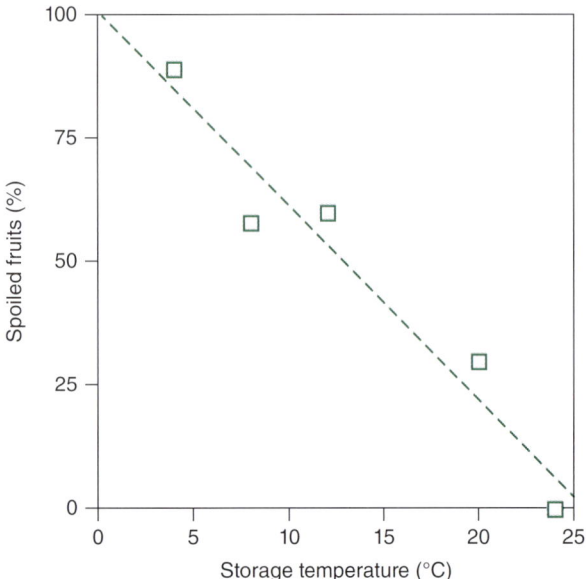

Fig. 7.1. Effect of storage temperature on the percentage of spoiled mature fruit of African horned melon after 37 days of storage. (Data from Benzioni *et al.*, 1993.)

Fig. 7.2. 'Golden Delicious' winter squash being transported from the field to a canning plant in Oregon. (Photo: Lynn Ketchum, OSU.)

extended storage, the bins are loaded on to trucks and moved into ventilated storage buildings by forklift.

Squash should be kept cool and relatively dry, preferably at 10–15°C and 50–70% RH. A higher RH of 70–80% will reduce shrinkage. For 'Butternut', weight loss should be kept below 15% to minimize development of a hollow neck. The disadvantage of high humidity storage is that microbial damage is greater. Hot-water treatment (2 min dip in 60°C water) has been used to reduce decay in 'Butternut' fruit stored at 70% RH (Francis and Thomson, 1965).

Mature squashes can be stored for 6 months or more, depending on the cultivar; *C. pepo* squashes such as acorn and delicata store well for 2–4 months, *C. moschata* squashes like butternut types are generally not of prime quality until after a month or two in storage and may be stored up to 5 months (Fig. 7.3). Longest storing are the *C. maxima* winter squashes, but this may vary among cultivars (Fig. 7.4). For example, 'Turban' will store for 3 months, 'Hubbard' for 6 months and 'Winter Sweet' and 'Sweet Meat' for 11 months. During this time, starch conversion continues and fruit colour and β-carotene content may improve. However, the palatability of 'Hubbard' fruit declines over time as overall carbohydrate content decreases and the percentage of water increases (Cummings and Jenkins, 1925).

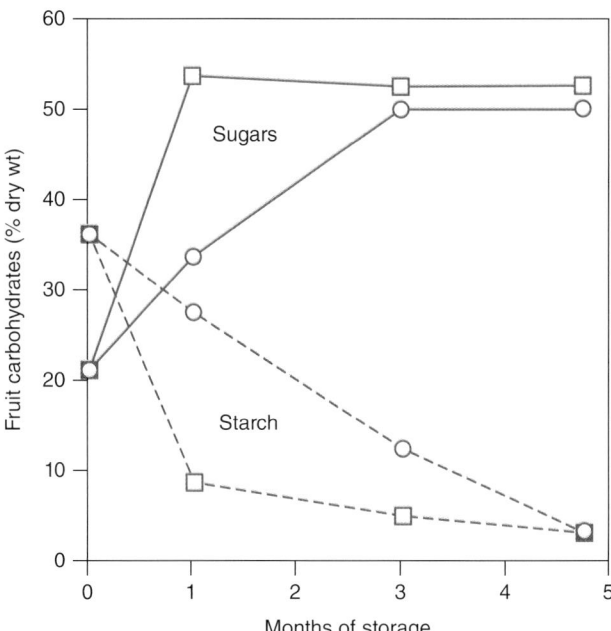

Fig. 7.3. Sugar (solid lines) and starch (dashed lines) content changes in fruit of 'Butternut' (*Cucurbita moschata*) during storage. Fruit were cured at about 26°C for 3 weeks (squares) or not cured (circles). (Data from Schales and Isenberg, 1963.)

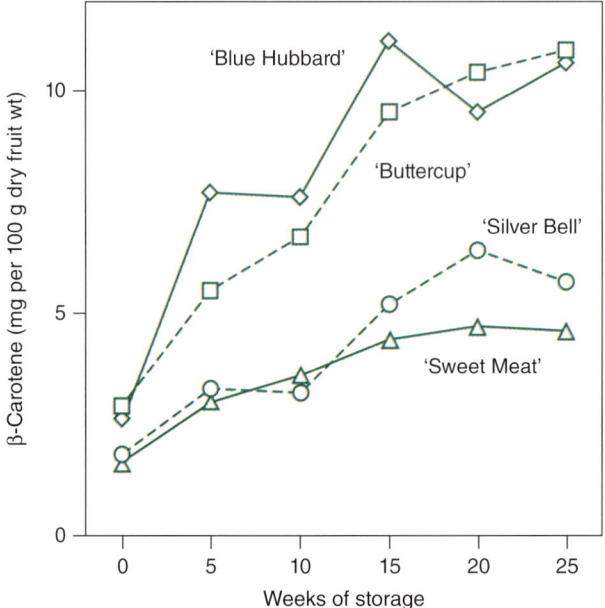

Fig. 7.4. Dry basis β-carotene content in mature fruit of four cultivars of winter squash (*Cucurbita maxima*) during cold storage. (Data from Hopp *et al.*, 1960.)

After harvest, melons are transported to a packing shed. If they retain much field heat, they are cooled in a hydrocooler, which is kept cold by a combination of ice and mechanical refrigeration. Some melons, particularly 'Crenshaw' and other thin-skinned cultivars, may be packed into padded shipping boxes in the field directly after they are picked. Melons transported long distances after harvest are often shipped in iced railroad cars or refrigerated trucks.

Melons can be stored for 1–6 weeks, depending on the cultivar and storage conditions. Cantalupensis Group melons fare well at 3–4°C and 85–90% RH for up to 2 weeks. Inodorus Group cultivars can be held for longer periods at temperatures up to 20°C.

'Honey Dew' melons are frequently treated with ethylene after shipment to promote uniform ripening of the harvest; however, immature fruit will not ripen off the vine satisfactorily, even with the application of ethylene. Application of 160 μl l⁻¹ ethylene for 24 h to African horned melons caused green, nearly mature fruit to turn yellow within 3 days of treatment compared with over 20 days for untreated fruit (Benzioni *et al.*, 1993). Atmospheric ethylene, from stored fruit such as apples, a faulty heater or other source, should be avoided when storing many other cucurbits; it causes undesirable yellowing of cucumbers and green-skinned squashes and flesh softening in watermelon.

Cucumbers should be handled carefully and cooled, in ways similar to melons, as soon as possible after harvest. Rough handling (e.g. dropped, smashed or cut fruit) during harvest or grading can cause bloating. Cultivars vary in their susceptibility to fruit damage. For example, fruit of 'Chipper' are more resistant to skin fracture than those of 'Pioneer'.

In the USA, growers take their crop to commercial receiving stations where cucumbers are sorted by grade and then trucked to pickle factories, where they may be stored before processing. At the pickling plant, cucumbers are: (i) brined, with or without fermentation (which can be controlled), and then stored or processed into dill, sweet or sour pickles; or (ii) packed with a cover liquor consisting of water, vinegar, salt and other flavourings, and then pasteurized to produce 'fresh pack' pickles or kept cold for refrigerated dills. These processes are illustrated in Fig. 7.5 and described in detail in Miller and Wehner (1989). For brined products, brine tanks are purged with air or nitrogen to remove CO_2 in order to reduce fruit bloating. Later, the pickles are desalted by leaching.

Recommended pre-processing storage conditions at pickling plants are 10°C and 95% RH (Etchells *et al.*, 1973). Lower levels of humidity cause fruit moisture loss and higher temperatures promote microbial growth. Cultivars vary in their storage life; under favourable conditions, 'Ohio MR200' and 'Marketer' will keep for 10 or 47 days, respectively. When recommended storage times are exceeded, mould and bacterial infections take hold (especially in spots where the spines have been broken off), the rind becomes loose on the shrivelling fruit and flesh flavour and texture deteriorate.

In experiments by Fellers and Pflug (1967), small (< 33 mm diameter) and large (33–51 mm) unwashed pickling cucumbers could be stored safely at 4.4°C for 6 and 9 days, respectively. Washing reduced storage life in half, whereas maintaining atmospheric levels of 5% CO_2 and 5% O_2 more than doubled storage life. However, such a low temperature is not usually employed in commercial operations, because chilling injury can occur and the fruit deteriorate quickly (within 18 h) after removal from storage.

As with picklers, high humidity (ca 95% RH) slows softening and shrivelling due to water loss in slicing cucumbers during storage. Slicers are waxed after harvest to keep the fruit from drying out as well as to enhance appearance. Glasshouse-grown cucumbers, which have a very tender skin, are often wrapped in plastic to prevent bruising and retard dehydration.

Watermelons should be harvested at maturity, since flavour and soluble solids content generally do not improve with storage. However, mature fruit are more easily injured during postharvest handling. They split easily if dropped and are subject to compression damage. They should be hand-packed in straw- or foam-padded crates, or into cardboard pallet boxes to prevent bruising and cracking during shipment.

Watermelons should be cooled as soon as possible after harvest. Delayed cooling and high humidity can lead to activation of anthracnose, a disease that

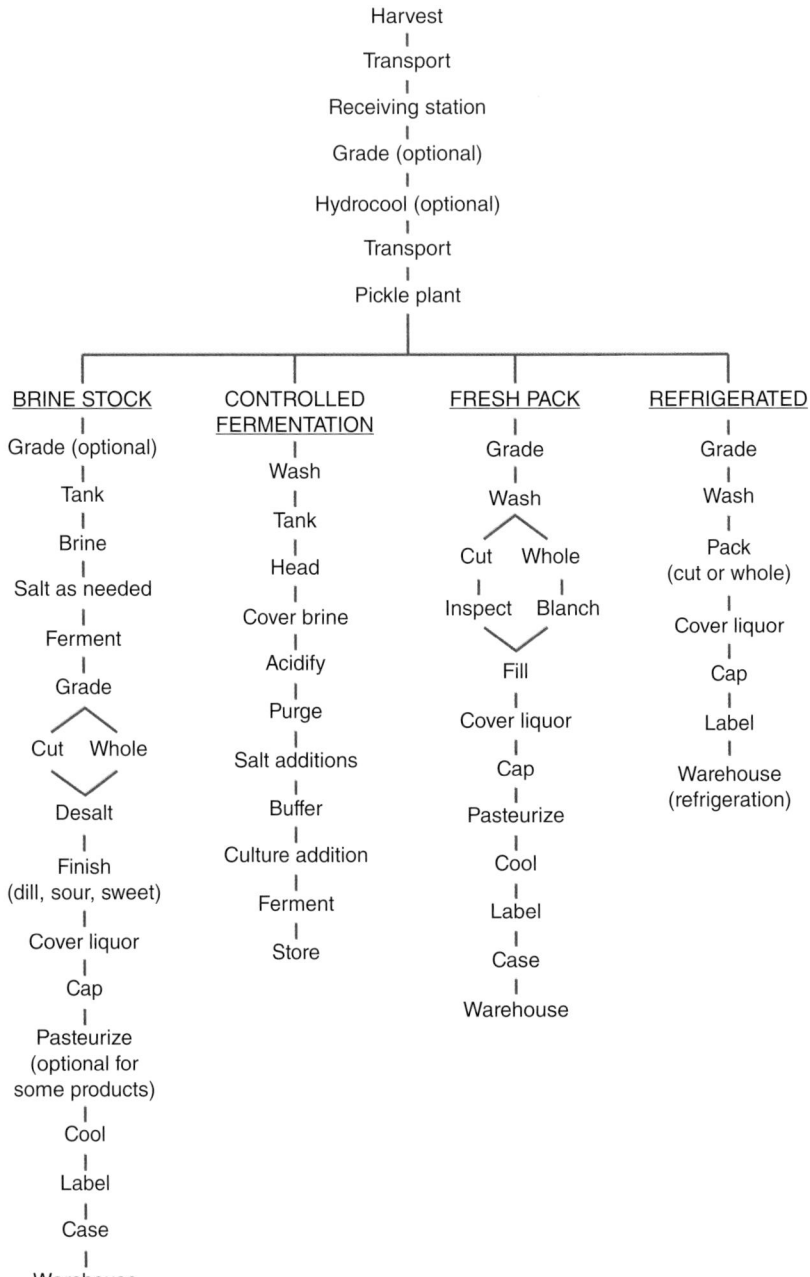

Fig. 7.5. Flow chart showing the movement of cucumbers from harvest to finished pickles. (Redrawn from Miller and Wehner, 1989. Reprinted with permission from N.A.M. Eskin (ed.) *Quality and Preservation of Vegetables*. Copyright CRC Press, Boca Raton, Florida, © 1989.)

may be latent in the field. Watermelons store satisfactorily at 15°C for up to 2 weeks. For long-term storage, the fruit should be kept at about 12°C and 85% RH; flesh colour fades below 10°C.

After mature bottle gourds are harvested, they need to dry out in a cool, dry, dark area for another 3–9 months, until the seeds rattle inside. Before storing, the freshly picked fruit should be washed thoroughly with a disinfectant solution to remove dirt and mould. During storage, they should be checked for mould weekly and cleaned if necessary. After curing is complete, the gourds are soaked to soften the outer skin, which can then be scraped off with a knife. After allowing the scraped fruit a day to dry, it is ready for crafting.

After harvest, mature luffa fruit grown for sponge production are usually soaked in warm water for about 15 min or until the skin slips off the fibrous interior. Most of the seeds can be removed beforehand by breaking the cap off at the blossom end of the fruit and shaking. With seeds, skin and excess pulp removed, the sponge is often rinsed in a bleach or hydrogen peroxide solution to whiten the fibres. Then the sponges are hung to dry.

Another cleaning method, which can produce better quality sponges, starts by harvesting the fruit just as it begins to turn brown. Squeezing the fruit causes the skin to pop loose from the fibres. Then the blossom end cap is broken off and a strip of rind is pulled down the length of the fruit to release the pulpy, fibrous interior. The sponge is rinsed clean of excess pulp and seeds and hung to dry.

Seeds

Cucurbit seeds will continue to develop even after the fruit is removed from the vine. If fruit are harvested before they are fully mature, due to impending frost or other reasons, it is advisable to store them for 1 or 2 months before extracting the seeds.

Whereas seeds of luffa and bottle gourd are extracted from dry fruit and require little or no processing before storage, mature seeds of most cucurbits need to be cleaned of wet, adherent flesh. If the seeds reside in the hollow cavity of the fruit (e.g. pumpkins and some melons), they need only be separated from the placental tissue, washed and dried. Embedded seeds of watermelon, cucumber, melon, wax melon and squash can be removed by chopping and smashing the fruit and adding water to the mixture; seeds will sink and the flesh debris, which floats, is poured off. For commercial seed harvesting, special machinery performs these operations in the field. Inventions include the seed sluice for small plots, the bulk seed extractor and the single-fruit seed extractor, which is used in the glasshouse or laboratory to extract cucumber seeds (Wehner and Humphries, 1995).

Removing the persistent placental material encasing cucumber, squash and bitter gourd seeds is aided by fermentation. The water and seed mixture is allowed to sit for 1–2 days, depending on the temperature (20–35°C).

Fermentation is complete when seeds settle to the bottom of the container and the placental material floats. The seeds are then rinsed and set out to dry.

A dilute solution of hydrochloric acid or ammonia can be used to clean cucumber seeds more quickly. After vigorous stirring, the seeds are free of flesh in 30 min or less. They must be rinsed before drying. This or any other mechanical means of seed cleaning should not be used with bitter gourd seeds, which are easily broken.

Cucurbit seeds are typically spread out to dry under warm (< 35°C) arid conditions. Commercial producers use forced air warmed by propane heaters and flat drying beds or large rotary dryers. Seeds ready for storage snap instead of bend and have a moisture content of about 5%. Dry seeds should be stored in dark airtight containers at about 5°C and 25% RH. Silica gel can be added to absorb moisture. Under these conditions, cucurbit seeds may remain viable for as long as 10 years or more.

The chayote seed can only be stored within the fleshy fruit. Mature fruit selected for propagation should be kept in a dark cool (ca 5°C) area for up to 6 months. Some shrivelling and decay of the fruit may occur during this time. When the next growing season arrives, the entire sprouted fruit is planted.

CUCURBITS FOR HUMAN HEALTH

INTRODUCTION

Cucurbit fruits are comparable in nutrient content to other fruits and vegetables and are exceptional for high pro-vitamin A content of some cultivars of squash and melon, and the abundant oil and protein in seeds of *Telfairia*, *Cucurbita* and other cucurbits. Cucurbits are a good source of carbohydrates and play a significant role in human nutrition, especially in tropical countries where their consumption is high.

There are three nutritional composition tables in this chapter: Tables 8.1 for raw cucurbits; Tables 8.2 for processed cucurbits; and Tables 8.3 for cucurbit seeds, flowers and leaves. Each table is divided into: (a) major components (moisture, calories, protein, lipids, fatty acids and minerals); and (b) vitamins and carotenoids.

On the basis of nutrients produced per worker-hour invested in production, MacGillivray *et al.* (1942) ranked *Cucurbita maxima* fruit first among all vegetable crops surveyed. This squash had the greatest amount of carotenoids produced per kilogram, per hectare and per worker-hour. In many nutrient comparisons, however, cucurbits were ranked low when compared with other vegetable crops. Cucumbers were ranked lowest in calories and thiamine, and watermelon lowest in protein and riboflavin per kilogram.

Leaves, stems and growing tips of cucurbits are eaten in some tropical countries. Cucurbit greens generally contain more calcium, phosphorus, ascorbic acid and iron than the fruit and can be good sources of pro-vitamin A also (Tables 8.1, 8.3). Leaves of various cultivated species contain up to 4–6% protein, while those of fluted pumpkin contain close to 2% fat and leaves of bitter gourd contain about 30 calories per 100 g.

Table 8.1a. Nutritional composition of raw cucurbits, per 100 g of fruit. Major components and including fatty acids and minerals[a].

Crop	Moisture (g)	Calories (Kcal)	Protein (g)	Lipid (g)	Fatty acids Saturated (g)	Fatty acids Mono-unsaturated (g)	Fatty acids Poly-unsaturated (g)	Carbohydrate (g)	Fibre (g)	Sugar (g)	Minerals Ca (mg)	Fe (mg)	Mg (mg)	P (mg)	K (mg)	Na (mg)	Zn (mg)	Cu (mg)	Mn (mg)	Se (µg)
Bitter gourd	94	17	1.00	0.17	–	–	–	3.70	2.8	–	19	0.43	17	31	296	5	0.80	0.034	0.089	0.2
Bottle gourd	96	14	0.62	0.02	0.002	0.004	0.009	3.39	0.5	–	26	0.20	11	13	150	2	0.70	0.026	0.066	0.2
Chayote	94	19	0.82	0.13	0.028	0.010	0.057	4.51	1.7	1.66	17	0.34	12	18	125	2	0.74	0.123	0.189	0.2
Cucumber	95	15	0.65	0.11	0.037	0.005	0.032	3.63	0.5	1.67	16	0.28	13	24	147	2	0.20	0.041	0.079	0.3
Horned melon (Kiwano)	89	44	1.78	1.26	–	–	–	7.56	–	–	13	1.13	40	37	123	2	0.48	0.020	0.039	–
Loofa	94	20	1.20	0.20	0.016	0.037	0.087	4.35	1.1	2.02	20	0.36	14	32	139	3	0.07	0.035	0.092	0.2
Melon, Cantaloupe	90	34	0.84	0.19	0.051	0.003	0.081	8.16	0.9	7.86	9	0.21	12	15	267	16	0.18	0.041	0.041	0.4
Melon, Casaba	92	28	1.11	0.10	0.025	0.002	0.039	6.58	0.9	5.69	11	0.34	11	5	182	9	0.07	0.060	0.035	0.4
Melon, Honeydew	90	36	0.54	0.14	0.038	0.003	0.059	9.09	0.8	8.12	6	0.17	10	11	228	18	0.09	0.024	0.027	0.7
Pumpkin	92	26	1.00	0.10	0.052	0.013	0.005	6.50	0.5	2.76	21	0.80	12	44	340	1	0.32	0.127	0.125	0.3
Summer squash, all types	95	16	1.21	0.18	0.044	0.016	0.089	3.35	1.1	2.20	15	0.35	17	38	262	2	0.29	0.051	0.175	0.2
Watermelon	91	30	0.61	0.15	0.016	0.037	0.050	7.55	0.4	6.20	7	0.24	10	11	112	1	0.10	0.042	0.038	0.4
Wax gourd	96	13	0.40	0.20	0.016	0.037	0.087	3.00	2.9	–	19	0.40	10	19	6	111	0.61	0.023	0.058	0.2
Winter squash, Acorn	88	40	0.80	0.10	0.021	0.007	0.042	10.42	1.5	–	33	0.70	32	36	347	3	0.13	0.065	0.167	0.5
Winter squash, Butternut	86	45	1.00	0.10	0.021	0.007	0.042	11.69	2.0	2.20	48	0.70	34	33	352	4	0.15	0.072	0.202	0.5
Winter squash, Hubbard	88	40	2.00	0.50	0.103	0.037	0.210	8.70	3.9	3.95	14	0.40	19	21	320	7	0.13	0.064	0.179	0.5
Winter squash, Spaghetti	92	31	0.64	0.57	0.117	0.042	0.239	6.91	1.5	2.76	23	0.31	12	12	108	17	0.19	0.037	0.125	0.3

[a]Data from US Department of Agriculture, Agricultural Research Service, Nutrient Data Laboratory. USDA National Nutrient Database for Standard Reference, Release 28. Version Current: September 2015, slightly revised May 2016. Internet: /nea/bhnrc/ndl.
– indicates not measured.

Table 8.1b Nutritional composition of raw cucurbits, per 100 g of fruit. Vitamins and carotenoids[a].

Crop	Ascorbic acid (mg)	Thiamin (mg)	Riboflavin (mg)	Niacin (mg)	Pantothenic Acid (mg)	B6 (mg)	Folate (µg)	Choline (mg)	E (mg)	K (µg)	A (IU[b])	A (RAE[b])	α-Carotene (µg)	β-Carotene (µg)	β-Cryptoxanthin (µg)	Lycopene (µg)	Lutein + Zeaxanthin (µg)
																	(Carotenoids)
Bitter gourd	84.0	0.040	0.040	0.40	0.21	0.04	72	–	–	–	471	24	185	190	–	–	170
Bottle gourd	10.1	0.029	0.022	0.32	0.15	0.04	6	–	–	–	16	–	–	–	–	–	–
Chayote	7.7	0.025	0.029	0.47	0.25	0.08	93	9.2	0.12	4.1	0	0	0	0	0	0	0
Cucumber	2.8	0.027	0.033	0.10	0.26	0.04	7	6.0	0.03	16.4	105	5	11	45	26	0	23
Horned melon (Kiwano)	5.3	0.025	0.015	0.57	0.18	0.06	3	–	–	–	147	7	–	88	–	–	–
Loofa	12.0	0.050	0.060	0.40	0.22	0.04	7	–	0.10	0.7	410	–	–	–	–	–	–
Melon, Cantaloupe	36.7	0.041	0.019	0.73	0.11	0.07	21	7.6	0.05	2.5	3,382	169	16	2,020	1	0	26
Melon, Casaba	21.8	0.015	0.031	0.23	0.08	0.16	8	7.6	0.05	2.5	0	0	0	0	0	0	26
Melon, Honeydew	18.0	0.038	0.012	0.42	0.16	0.09	19	7.6	0.02	2.9	50	3	0	30	0	0	27
Pumpkin	9.0	0.050	0.110	0.60	0.30	0.06	16	8.2	1.06	1.1	8,513	426	4,016	3,100	0	0	1,500
Summer squash, all types	17.0	0.048	0.142	0.49	0.16	0.22	29	6.7	0.12	3.0	200	10	0	120	0	0	2,125
Watermelon	8.1	0.033	0.021	0.18	0.22	0.05	3	4.1	0.05	0.1	569	28	0	303	78	4,532	8
Wax gourd	13.0	0.040	0.110	0.40	0.13	0.04	5	–	–	–	0	0	–	–	–	–	–
Winter squash, Acorn	11.0	0.140	0.010	0.70	0.40	0.15	17	–	–	–	367	18	0	220	0	0	38
Winter squash, Butternut	21.0	0.100	0.020	1.20	0.40	0.15	27	11.8	1.44	1.1	10,630	532	834	4,226	3471	0	0
Winter squash, Hubbard	11.0	0.070	0.040	0.50	0.40	0.15	16	–	0.16	1.3	1,367	68	0	820	0	0	0
Winter squash, Spaghetti	2.1	0.037	0.018	0.95	0.36	0.10	12	8.2	0.13	0.9	120	6	16	64	0	0	0

[a]Data from US Department of Agriculture, Agricultural Research Service, Nutrient Data Laboratory. USDA National Nutrient Database for Standard Reference, Release 28. Version Current: September 2015, slightly revised May 2016. Internet: /nea/bhnrc/ndl.

[b]IU = international units; RAE = Retinol Activity Equivalents.

– indicates not measured.

Table 8.2a. Nutritional composition of processed (dried, pickled, canned or frozen) cucurbits, per 100 g fruit. Major components and including fatty acids and minerals[a].

Crop	Preparation[b]	Moisture (g)	Calories (kcal)	Protein (g)	Lipid (g)	Saturated (g)	Mono-unsaturated (g)	Poly-unsaturated (g)	Carbohydrate (g)	Fibre (g)	Sugar (g)	Ca (mg)	Fe (mg)	Mg (mg)	P (mg)	K (mg)	Na (mg)	Zn (mg)	Cu (mg)	Mn (mg)	Se (µg)
																Fatty acids				Minerals	
Bottle Gourd, (Kanpyo)	Dried strips	20	258	8.58	0.56	0.045	0.103	0.244	65.03	9.8	–[b]	280	5.12	125	188	1582	15	5.86	0.433	1.137	2.6
Melon balls	Frozen	90	33	0.84	0.25	0.064	0.006	0.098	7.94	0.7	–	10	0.29	14	12	280	31	0.17	0.060	0.040	–
Pickles, Cucumber	Dill or Kosher	94	12	0.50	0.30	0.079	0.005	0.122	2.41	1.0	1.07	57	0.26	7	16	117	809	0.10	0.028	0.062	0
Pickles, Cucumber	Sour	94	11	0.33	0.20	0.052	0.003	0.081	2.26	1.2	1.06	0	0.40	4	14	23	1,208	0.02	0.085	0.011	0
Pickles, Cucumber	Sweet	76	91	0.58	0.41	0.067	0.004	0.106	21.15	1.0	18.27	61	0.25	7	18	100	457	0.12	0.028	0.065	0
Pumpkin	Canned	90	34	1.10	0.28	0.146	0.037	0.015	8.09	2.9	3.30	26	1.39	23	35	206	5	0.17	0.107	0.149	0.4
Summer squash	Canned, drained	96	13	0.61	0.07	0.015	0.005	0.031	2.96	1.4	1.19	12	0.71	13	21	96	5	0.29	0.080	0.097	0.2
Summer squash	Frozen	94	20	0.83	0.14	0.029	0.011	0.060	4.80	1.2	–	18	0.48	23	35	209	5	0.37	0.085	0.162	0.2
Winter squash	Baby food	93	28	0.81	0.20	0.020	0.020	0.040	5.73	0.9	3.37	24	0.32	14	21	185	5	0.19	0.030	0.083	0.2
Winter squash, Butternut	Frozen	83	57	1.76	0.10	0.021	0.007	0.042	14.41	1.3	2.83	29	0.88	14	22	212	2	0.17	0.051	0.248	0.7

[a]Data from US Department of Agriculture, Agricultural Research Service, Nutrient Data Laboratory. USDA National Nutrient Database for Standard Reference, Release 28. Version Current: September 2015, slightly revised May 2016. Internet: /nea/bhnrc/ndl.
[b]Prepared without added salt and sugar except pickles.
– indicates not measured.

Table 8.2b Nutritional composition of processed (dried, pickled, canned or frozen) cucurbits, per 100 g fruit. Vitamins and carotenoids[a].

Crop	Preparation[b]	Vitamins												Carotenoids			
		Ascorbic acid (mg)	Thiamin (mg)	Riboflavin (mg)	Niacin (mg)	Pantothenic Acid (mg)	B6 (mg)	Folate (µg)	Choline (mg)	E (mg)	K (µg)	A (IU[2])	A (RAE[2])	α-Carotene (µg)	β-Carotene (µg)	β-Cryptoxanthin (µg)	Lutein + Zeaxanthin (µg)
Bottle Gourd, (Kanpyo)	Dried strips	0.2	0.000	0.044	2.90	2.55	0.53	61	–	–	–	0	–	–	–	–	–
Melon balls	Frozen	6.2	0.166	0.022	0.64	0.16	0.11	26	–	–	–	1,774	89	–	–	–	–
Pickles, Cucumber	Dill or Kosher	2.3	0.045	0.057	0.11	0.20	0.04	8	3.4	0.03	17.3	125	6	13	53	30	28
Pickles, Cucumber	Sour	1.0	0.000	0.010	0.00	0.04	0.01	1	3.6	0.09	47.0	191	10	20	81	47	43
Pickles, Cucumber	Sweet	0.7	0.025	0.030	0.12	0.05	0.02	1	3.1	0.36	47.1	764	38	81	325	186	170
Pumpkin	Canned	4.2	0.024	0.054	0.37	0.40	0.06	12	9.8	1.06	16.0	15,563	778	4,795	6,940	0	0
Summer squash	Canned, drained	2.7	0.016	0.027	0.42	0.05	0.04	10	4.9	0.12	2.8	102	5	0	61	0	198
Summer squash	Frozen	6.4	0.040	0.048	0.40	0.09	0.09	12	–	–	–	279	14	–	–	–	–
Winter squash	Baby food	0.3	0.029	0.048	0.68	0.21	0.07	8	7.6	0.55	4.7	1,703	85	242	896	9	3,527
Winter squash, Butternut	Frozen	6.2	0.090	0.059	0.74	0.22	0.11	24	–	1.85	1.4	4,790	240	376	1,904	1564	0

[a]Data from US Department of Agriculture, Agricultural Research Service, Nutrient Data Laboratory. USDA National Nutrient Database for Standard Reference, Release 28. Version Current: September 2015, slightly revised May 2016. Internet: /nea/bhnrc/ndl.

[b]Prepared without added salt and sugar except pickles.

[2]IU = international units; RAE = Retinol Activity Equivalents.

– indicates not measured.

Table 8.3a. Nutritional composition of cucurbit seeds, flowers and leaves per 100 g of material. Crops reported here are raw or minimally processed, without salt. Major components and including fatty acids and minerals[a].

Crop	Plant part & preparation	Moisture (g)	Calories (Kcal)	Protein (g)	Lipid (g)	Fatty acids			Carbohydrate (g)	Fibre (g)	Sugar (g)	Minerals									
						Saturated (g)	Mono-unsaturated (g)	Poly-unsaturated (g)				Ca (mg)	Fe (mg)	Mg (mg)	P (mg)	K (g)	Na (mg)	Zn (mg)	Cu (mg)	Mn (mg)	Se (µg)
Seeds																					
Watermelon	Kernels, dried	5	557	28.3	47.4	9.78	7.41	28.09	15.3	–	–	54	7.3	515	755	648	99	10.2	0.69	1.61	–
Pumpkin	Kernels, dried	5	559	30.2	49.1	8.66	16.24	20.98	10.7	6.0	1.4	46	8.8	592	1,233	809	7	7.8	1.34	4.54	9.4
Pumpkin	Seeds, whole roasted	5	446	18.6	19.4	3.67	6.03	8.84	53.8	18.4	–	55	3.3	262	92	919	18	10.3	0.69	0.50	–
Flowers & leaves																					
Bitter gourd	Leafy tips	89	30	5.3	0.7	–	–	–	3.3	–	–	84	2.0	85	99	608	11	0.3	0.20	0.54	0.9
Pumpkin	Flowers	95	15	1.0	0.1	0.04	0.01	0.00	3.3	–	–	39	0.7	24	49	173	5	–	–	–	0.7
Pumpkin	Leaves	93	19	3.2	0.4	0.21	0.05	0.02	2.3	–	–	39	2.2	38	104	436	11	0.2	0.13	0.36	0.9

[a]Data from US Department of Agriculture, Agricultural Research Service, Nutrient Data Laboratory. USDA National Nutrient Database for Standard Reference, Release 28. Version Current: September 2015, slightly revised May 2016. Internet: /nea/bhnrc/ndl.
– indicates not measured.

Table 8.3b. Nutritional composition of cucurbit seeds, flowers and leaves per 100 g of material. Crops reported here are raw or minimally processed, without salt. Vitamins and carotenoids[a].

Crop	Plant part & preparation	Ascorbic acid (mg)	Thiamin (mg)	Riboflavin (mg)	Niacin (mg)	Pantothenic Acid (mg)	B6 (mg)	Folate (µg)	Choline (mg)	E (mg)	K (µg)	A (IU[b])	A (RAE[b])	α-Carotene (µg)	β-Carotene (µg)	β-Cryptoxanthin (µg)	Lutein + Zeaxanthin (µg)
					Vitamins										Carotenoids		
Seeds																	
Watermelon	Kernels, dried	0.0	0.19	0.15	3.55	0.35	0.09	58	–	–	–	0	0	–	–	–	–
Pumpkin	Kernels, dried	1.9	0.27	0.15	4.99	0.75	0.14	58	63	2.2	7.3	16	1	1	9	1	74
Pumpkin	Seeds, whole roasted	0.3	0.03	0.05	0.29	0.06	0.04	9	–	–	–	62	3	–	–	–	–
Flowers & leaves																	
Bitter gourd	Leafy tips	88.0	0.18	0.36	1.11	0.06	0.80	128	–	–	–	1,734	87	–	–	–	–
Pumpkin	Flowers	28.0	0.04	0.08	0.69	–	–	59	–	–	–	1,947	97	–	–	–	–
Pumpkin	Leaves	11.0	0.09	0.13	0.92	0.04	0.21	36	–	–	–	1,942	97	–	–	–	–

[a]Data from US Department of Agriculture, Agricultural Research Service, Nutrient Data Laboratory. USDA National Nutrient Database for Standard Reference, Release 28. Version Current: September 2015, slightly revised May 2016. Internet: /nea/bhnrc/ndl.

[b]IU = international units; RAE = Retinol Activity Equivalents.

– indicates not measured.

MOISTURE AND SOLIDS CONTENT IN FRUIT

Fruit of most cucurbits have a high moisture content (Tables 8.1, 8.2). Inhabitants of the Kalahari Desert in Africa and some other arid areas consume watermelons as a source of water. Cucumbers have 95% moisture content, higher than that for most other vegetables, and they have little nutritional value. Wax gourd fruit also are high in moisture content and low in nutrients. In contrast, baked winter squash has less moisture content (80–85%) than most other cooked vegetables (Tables 8.1a, 8.2a).

Composition of pickled cucumbers varies according to the processing method (Tables 8.2a, b). Dill pickles and sour pickles are low in caloric content, whereas sweet pickles and relishes have more calories due to the sugar added during processing. Pickles brined during fermentation have a high sodium content.

CARBOHYDRATES

Carbohydrates enter the developing fruit from leaves in a soluble form, often as stachyose rather than sucrose, and then the sugars are converted into reducing sugars (e.g. glucose and fructose). In melons, sucrose accumulates later in the maturation process, with the percentage of sucrose produced increasing with fruit maturity (Fig. 8.1). Sugars are important quality components for squash, melon, watermelon and many other cucurbit fruit. Their content can easily be determined with a hand refractometer. Honeydew melons generally have a higher sugar content (10–16%) than Cantalupensis Group melons (8–14%). Long shelf-life (LSL) melons and newer higher sugar melons, including Galia melons, have much higher soluble solids content. In squash, stachyose is the initial substrate for starch biogenesis and soluble sugars remain low until rather late in fruit development. Starch degradation varies widely among cultigens, but tends to be most rapid in acorn and later in *Cucurbita moschata* cultigens. Starch conversion, which accounts for greater sugar content in older squashes, continues in mature fruit under storage (Culpepper and Moon, 1945). In African horned melon, sugar content increases during fruit storage, but is even higher if fruit are allowed to ripen on the vine, peaking at 50 days after pollination (Benzioni *et al.*, 1993). During cucumber development, sugar content (mostly glucose and fructose) follows a sigmoidal curve; increase in the parthenocarpic cultivar 'Brilliant' was rapid for the first 6 days following flowering and then slowed until a maximum was reached near the beginning of senescence (Davies and Kempton, 1976).

Among squash species, *C. maxima* and *C. moschata* produce fruit with the strongest taste, highest solids and deepest flesh colour. Therefore, they are preferred for commercial canning. *C. maxima* types tend to be higher in total solids (approximately 12%) than *C. moschata* (4.5–6%), but the latter may have

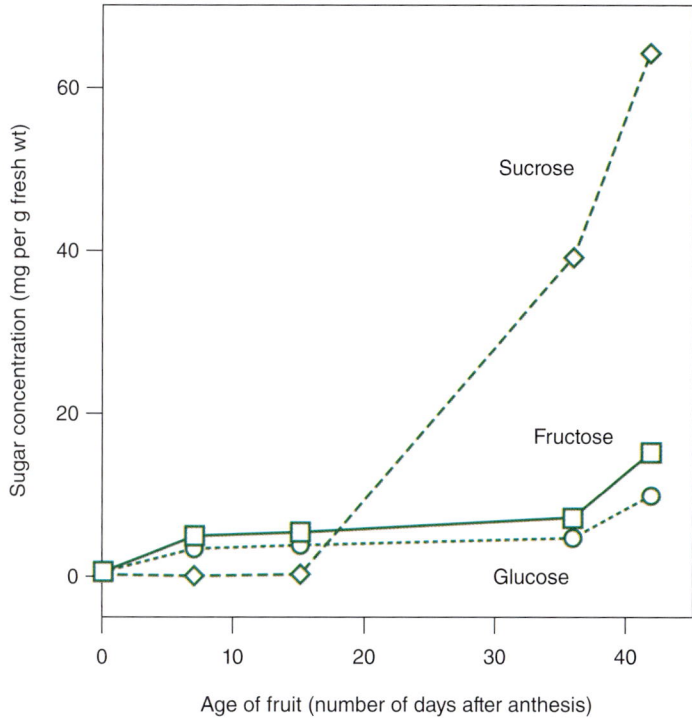

Fig. 8.1. Concentrations of the primary sugars in developing fruit of melon. (Data from Hughes and Yamaguchi, 1983.)

higher yields. Another consideration is that some *C. maxima* cultivars, such as 'Golden Delicious', are used for both processing and culinary seed. Consistency, an important consideration for canning, refers to the stiffness of the processed flesh, which is greatly influenced by starch and soluble solids content. Because the conversion of starch to sugars increases during fruit storage, freshly harvested fruit are chosen for processing.

Chayote roots have been nutritionally analysed and are a good source of easily digestible starch. Research on buffalo gourd roots revealed a dry-weight starch content of 55% and a digestibility of 41% when consumed raw and 95% when cooked. Once the cucurbitacins are removed through filtering, processed buffalo gourd starch compares favourably with corn starch in a variety of food industry applications (Gathman and Bemis, 1990).

PROTEINS AND OILS

Seeds, the most nutritious part of the cucurbit plant, are often thrown away. Seeds discarded in most countries are eaten in some countries, e.g. watermelon seeds in China and squash (pumpkin) seeds in Mexico. Some cucurbit species, including oyster nut, are grown primarily for their edible seeds. For some crops, cultivars have been bred specifically for seeds, such as citron, egusi and Styrian oilseed pumpkins. Seeds from squash, pumpkin, melon, watermelon and other mature cucurbit fruits are good sources of protein, calcium, phosphorus, iron, magnesium, arginine, methionine, aspartic acid, glutamic acid, thiamin and niacin (Tables 8.3a, b). Decorticated, or shelled, seeds are rich in fats (40–60%) and proteins (30–40%), have approximately threefold higher levels of potassium compared with flesh and contain few free sugars. As in other oilseed crops, cucurbit seed proteins, which consist of mostly globulins and some albumins, have abundant nitrogen-rich amino acids and are deficient in lysine and sulfur-containing amino acids.

Cucurbit seed oils, which are usually edible and suitable for various industrial purposes, are composed primarily of unsaturated fatty acids (Table 8.3a). The semi-drying oils of squash, melon, watermelon and luffa seeds contain 40–60% linoleic acid, 20–40% oleic acid, 10–20% palmitic acid and 0–15% stearic acid. Cucumber seeds have about 59% oleic, 22% linoleic, 7% palmitic and 4% stearic acids. With an overall flavour of groundnuts, oil in seeds of white-seeded melon is highest in linolenic acid (65%). Seeds of fluted pumpkin, which taste like almonds, are very rich in an oil composed of 37% oleic, 21% palmitic, 21% stearic and 15% linoleic acids. Breeding efforts with buffalo gourd have produced seeds with linoleic acid content ranging up to 82%. However, environmental growing conditions greatly affect the percentage of unsaturated fats.

VITAMINS, MINERALS AND SECONDARY COMPOUNDS

Carotenoids

Cucurbit fruits with orange, red or yellow flesh generally have high concentrations of carotenoids. Carotenoids can be divided into two basic classes: (i) hydrocarbon carotenes; and (ii) oxygenated xanthophylls. Some carotenes, such as β-carotene, are the precursors of vitamin A, an essential vitamin to humans. Consumption of carotenoids is associated with reduced incidence of cancers and cardiovascular disease and certain ones (lutein and zeaxanthin) are associated with preventing macular degeneration. In breeding programmes, it is ideal to have chemical analyses for vitamins and nutrients. If that is not possible, the breeder can still make significant improvement in carotenoid content simply by selecting for deep orange flesh colour. One approach is to use the Roche Yolk Color Fan to select flesh colours high in carotene pigments, which are under dominant genetic control in some varieties of *C. moschata*. Many of the early studies assumed that most of the carotenoids in squash were carotenes and thus promoted squash as a source of pro-vitamin A.

However, more recent studies of carotenoid profiles in squash demonstrated extreme variation in carotenoid profiles (Azevedo-Meleiro and Rodriguez-Amaya, 2007; Obrero *et al.*, 2013; Durante *et al.*, 2014; Bonina-Noseworthy *et al.*, 2016). Because the flesh in most squash cultivars of *C. maxima* and *C. moschata* has moderately high contents of lutein and may also have fairly high contents of β-carotene, these types may have a dual role in promoting human health. The red and orange coloration in watermelon cultivars is due to the non-provitamin carotenes lycopene and prolycopene.

Carotenes

Carotenes found in cucurbits include α-, β-, δ- γ- and ζ-carotene. Some of these are converted in the human small intestine into retinal (vitamin A) with efficiency of the process varying with the type of carotene. β-Carotene is the most efficient, with symmetrical cleavage of the molecule producing two retinyl groups. The α- and γ-carotenes have only a single retinyl group, while others (δ- and ζ-carotene) lack retinyl groups entirely.

Xishuangbanna gourd from China has orange flesh and more β-carotene than other cucumber cultivars. Navazio (1994), who studied carotenoid content and inheritance in these cucumbers, found that carotene content increases with fruit age, is affected by growing conditions and is controlled by at least two genes. In orange-fleshed watermelon, prolycopene, phytoene and ζ-carotene are the major carotenoids. Orange-fleshed melons and squashes are much better sources of β-carotene than cucumbers (Table 8.1b). 'Honey Dew', 'Casaba' and other green- or white-fleshed melons have lower carotene compositions than melon cultivars with orange flesh.

Lycopene is a red-pigmented carotene with antioxidant properties that serves as an intermediate for the biosynthesis of other carotenoids. In red-fleshed watermelon, lycopene accounts for 70–90% of the total carotenoids, the remainder consisting of phytofluene, phytoene, β-carotene, lutein, nerosporene, and ζ-carotene. Studies have shown that the content of lycopene and carotenoids increases rapidly and accumulates 10–12 days after pollination in diploid watermelons, and continues to accumulate as the fruit mature.

In humans, lycopene reduces cancer cell growth and induces cell death in malignant leukaemia and in endometrial, mammary, lung and prostate cancer cells. Lycopene protects against lipid peroxidation and foam cell production, which are implicated in the initiation of atherosclerosis. A diet consisting of fruits and vegetables rich in lycopene can protect against stroke and cardiovascular diseases. Watermelon juice containing lycopene and citrulline may improve athlete recovery and performance.

Xanthophylls

Xanthophylls are also important, especially in squash and pumpkins (Tables 8.1b, 8.2b). In the older literature, xanthophylls and carotenes were often combined, resulting in inflated estimates of carotenes and pro-vitamin

A activity. Forms most common in cucurbits include lutein, zeaxanthin, neoxanthin, violaxanthin, neochrome and β-cryptoxanthin. Only the latter has pro-vitamin A activity, but lutein and zeaxanthin are implicated in eye health. The predominant xanthophyll found in squash is lutein. While zeaxanthin has been reported in the literature, it has been difficult to verify the presence of this compound. Lutein is a yellow carotenoid that is often found in the yellow or dark green vegetables, including zucchini squash ($2125\ \mu g\ 100g^{-1}$) and cucumber ($23\ \mu g\ 100g^{-1}$), but it reaches highest levels in processed winter squash ($3527\ \mu g\ 100g^{-1}$) (Table 8.2b). In humans, lutein protects the eye from oxidative stress. In canary yellow and salmon yellow fleshed watermelons, neoxanthin is the major carotenoid (Bang *et al.*, 2010).

Citrulline

Citrulline is present in some cucurbits, including watermelon. It belongs to the L-arginine metabolic family. In related *Citrullus* species, citrulline content increases in the foliage during drought stress and may improve plant tolerance to stress. In melon, foliage citrulline content is an important indicator for drought stress. Watermelon is a good source of arginine and the arginine precursor, citrulline. In developing watermelon fruit, citrulline content is low, reaching peak levels just before maturity and declining as the fruit age. Citrulline content tends to be higher in canary yellow watermelons.

In humans, citrulline helps in muscle recovery during exercise and benefits vascular health, such as reducing blood pressure and increasing vasodilation in many tissues of the body, which helps to mitigate cardiovascular disease.

Flavonoids and phenolic acids

Phenolic acids and flavonoids are water-soluble compounds recognized as having health benefits and, in some cases, important pigmentation properties. Cucurbits tend not to be high in these compounds, with low quantities of hydroxybenzoic and hydroxycinnamic acids reported in cucumber, melon and pumpkin (Macheix *et al.*, 2000). In terms of flavonoids, flavan-3-ols (including epicatechin and epigallocatechin), flavones (apigenin and luteolin), flavonols (kaempferol, quercetin and rutin) have been detected (USDA Nutrient Database, 2014). Flavonoid classes that have not been detected in cucurbits include flavanones and anthocyanins, the latter with red to blue pigment properties. In many fruit crops, including cucurbits, phenolic acids and flavonoids are found predominantly in the epidermis and outer layers of the pericarp and may be lost when the skin of the fruit is discarded.

ANTINUTRIENTS AND PHARMACOLOGY

Toxic and potentially therapeutic compounds in cucurbits include oxygenated tetracyclic triterpenoids called cucurbitacins, saponins (e.g. cucurbitocitrin in watermelon seeds, tubeimoside 1 in pseudo-fritillary), other glycosides (citrullol and colocynth in *Citrullus colocynthis*), alkaloids (momordicin in bitter gourd), ribosome-inactivating proteins (luffaculin in *Luffa operculata*, trichosanthin in *Trichosanthes*), free amino acids (cucurbitin in squash) and various other compounds. One interesting glycoside is mogrol I-IV; found in fruit of luo-han-guo, it is 150 times sweeter than sucrose and is being investigated as a natural, non-calorie sugar substitute.

The most intensely studied of these biodynamic compounds are the bitter cucurbitacins, because of their unwanted presence in fruit of the major cucurbit crops. Cucurbitacins in cultivated cucurbits can be a serious problem, not just due to the extremely unpleasant flavour they impart, but also because of health concerns. These compounds are potent purgatives and laxatives and in high concentrations can cause serious illness and death.

Cucurbitacins commonly occur in the Cucurbitaceae, hence their name, as well as in a few species of other families. There are many different cucurbitacin compounds, existing as glycosides or free aglycones. Those in squash are particularly toxic. However, the highest concentrations (> 1%) are found in the fruit of colocynth and various wild species of *Cucumis*. Within a particular plant, concentrations are usually highest in the fruit and roots and lower in leaves, stems and growing tips, but concentrations may vary in a plant part during development.

Different letters of the alphabet have been assigned to different cucurbitacin compounds. These distinct but structurally related compounds have some chemotaxonomic value, since some species and groups of species can be characterized by the cucurbitacins they contain. Thus, while only cucumber has cucurbitacin C, squash species are characterized by B, D, E, I and the glycoside of E.

The bitter, toxic cucurbitacins were of selective value during cucurbit evolution, by deterring insects and other herbivores from consuming foliage and fruit. Aphids, spider mites and other pests are repelled by cucurbitacins, but cucumber beetles are not. During the coevolution of plants and insects, cucumber beetles developed an attraction to plants with high concentrations of cucurbitacins.

Wild species of *Cucurbita* have bitter fruit and a key step in squash domestication was selection for types with non-bitter fruit. The first plant part consumed by people was probably the seed, which is non-bitter even when fruit flesh is bitter. The cucurbitacins are concentrated in the placenta, the tissue attached to the seed, and the seeds are not bitter if thoroughly cleaned and freed of placenta and adhering flesh. During the process of extracting seeds, bitterness from the flesh would have transferred to the fingers of early cultivators.

Consequently, mutants with non-bitter flesh would probably have been noticed and their seeds saved for planting. In this way, non-bitter squash was probably selected for and cultivated by Native Americans thousands of years ago.

Cucurbitacins occur in the fruit of all squash cultivars, but normally they are in such low concentrations that they cannot be tasted. Occasionally, however, relatively high concentrations make a squash quite bitter. There have been a number of cases where people became very ill, requiring hospitalization after eating only a few grams of bitter-tasting fruit.

Bitterness in cucurbits is influenced by both genetic and environmental factors. At least five genes regulate cucurbitacin biosynthesis, with organ-specific genes controlling the quality and quantity of cucurbitacins in different plant parts. In squash, problems encountered with bitterness are believed to be primarily genetic in origin. Many ornamental gourds, as well as wild populations of *Cucurbita pepo*, have a single dominant allele for bitter fruit. This allele can be transferred to *C. pepo* squash cultivars by pollinating insects, and plants of future generations may then have bitter fruit. There is no xenia for fruit bitterness; if squash crosses with a gourd or another source of the allele for bitter fruit, the bitterness is not expressed until the F_1 generation.

Although bitterness is most frequently a problem with *C. pepo* cultivars, probably as a result of gene flow from bitter cultivars or wild populations, it has also been reported in the other squash species. As a result of gene interaction, bitter fruit can develop in the progeny of *C. pepo* × *C. argyrosperma*, even when both parents have non-bitter fruit.

In cucumber, environmental factors often influence when bitterness occurs. One fruit may be bitter, while another fruit on the same plant, developing under different weather conditions, may be non-bitter. Some cucumber cultivars have bitter foliage but never have bitter fruit; others have bitter foliage and may have bitter or non-bitter fruit, depending on environmental conditions. It is difficult to distinguish between these types in a breeding programme because of the environmentally induced variability. However, Dutch breeders developed a better method of breeding cucumbers for non-bitter fruit. Andeweg and Bruyn (1958) reasoned that if they could find a cucumber mutant with non-bitter foliage, it would have non-bitter fruit regardless of the environment and growing conditions. After tasting the cotyledons of 15,000 cucumber seedlings, these researchers found one that was non-bitter. Upon growing the seedling out, it produced non-bitter fruit under all growing conditions. The single recessive allele involved has been bred into many modern cultivars.

9

DISEASES AND NEMATODES

INTRODUCTION

Cucurbits suffer from many diseases that are caused by bacteria, fungi (and oomycetes), nematodes, viruses and phytoplasmas (Blancard *et al.*, 1994; Zitter *et al.*, 1996). Disease pathogens attack cucurbits at every stage of development, from damping off at the beginning of germination to postharvest fruit rots. Reliable fruit yield and quality of cucurbits is dependent on adequate pathogen control.

Disease control is easiest for growers if done with genetically resistant crops. The first crop ever bred for disease resistance was a cucurbit: a watermelon cultivar ('Conqueror') resistant to Fusarium wilt was released by the USDA in 1911. Watermelon cultivars are available with resistance to anthracnose race 1 and Fusarium wilt race 1. Melon cultivars are available with resistance to sulfur burn, melon necrotic spot virus, powdery mildew and Fusarium wilt races 0, 1 and 2. Much progress has been made in breeding cucumber for disease resistance. Sources of resistance have been found for most of the major pathogens of cucumber, and scientists have developed germplasm with multiple disease resistance. For example, 'Wisconsin 2757' is resistant to scab, cucumber mosaic virus (CMV), watermelon mosaic virus (WMV), bacterial wilt, angular leaf spot, anthracnose, powdery mildew, downy mildew, target leaf spot and Fusarium wilt. Squash cultivars are available with resistance to virus and powdery mildew, with some having resistance to papaya ringspot virus (PRSV), watermelon mosaic virus and zucchini yellow mosaic virus (ZYMV). Squash was one of the first crops (approved in 1995) to have traits (virus resistance from virus coat protein genes) from transformation using *Agrobacterium tumefaciens*, making them genetically modified organisms (GMO).

Where breeding has not yet produced disease resistance, various preventive and control measures are needed to reduce crop losses. Phytosanitary procedures, including crop rotation and seed treatment, are employed to avoid infection of susceptible plants. Growers often apply chemicals (e.g. fungicides, bactericides, nematicides) preventively as well. Insecticides and other means are used to control the insect vectors carrying contagions, especially viruses. Some susceptible cultivars are grafted to the rootstocks of disease-resistant cucurbits (see 'Grafting' in Chapter 6 for more details).

Early disease detection (before visible signs appear) is possible using enzyme-linked immunosorbent assay (ELISA) kits, digital droplet PCR, LAMP (loop-mediated isothermal amplification) and other diagnostic assays. In ELISA the presence of the infectious agent in the plant is revealed by testing the sap for pathogen-specific proteins. Some kits can be operated in the field to test for various species of fungi, with results in 10 min.

Disease development is often mediated by environmental conditions. For example, ultraviolet-B radiation treatment can increase fungal susceptibility in some cucumber cultivars (Orth *et al.*, 1990). However, UV-B radiation can also harm the fungus and impair disease progression. Fusarium wilt infection of watermelon roots is promoted by low soil pH, low to moderate soil moisture (25% saturation), and the use of NH_4 instead of NO_3 fertilizer. Sudden wilt of melon can be caused by a variety of pathogens in combination with certain environmental conditions, for example when cool rainy weather is followed by hot sunny weather during the fruiting season. In general, environmentally stressed cucurbits are more susceptible to initial infection and subsequent disease development.

BACTERIAL AND PHYTOPLASMA DISEASES

Angular leaf spot

Angular leaf spot is caused by a bacterium (*Pseudomonas syringae* pv. *lachrymans*) that is distributed worldwide. *Pseudomonas marginalis* and *P. viridiflava* have also been reported to produce a bacterial wilt of cucumber in Japan. Angular leaf spot is primarily a disease of cucumber (Fig. 9.1), but it damages watermelon, squash and melon as well.

Fig. 9.1. Angular lesions on cucumber from angular leaf spot. (Photo: J.R. Myers.)

Infection occurs through wounds and stomata. Water-soaked spots develop into small, angular lesions on the foliage, flowers and fruit. The leaf lesions may produce a white, milky exudate on the lower leaf surface before dropping out, leaving holes up to 30 mm in diameter. The nearly circular fruit lesions, smaller than those on the leaf, become white to tan and often crack. An infected fruit may rot due to secondary soft rot bacteria entering the fruit through these lesions.

The bacterium overwinters in the soil or on plant debris and can also reside on the seed surface or below the seed coat. Since it is seed transmitted, the use of disease-free seeds or seed treatment with hot, acidic preparations is recommended. Seeds should be produced where the disease does not occur, with at least a 2-year period between cucurbit crops in a field. The pathogen is also disseminated by the cucurbit leaf beetle (*Aulacophora femoralis*) and control of this insect helps to contain angular leaf spot. Keeping the plants dry, by avoiding excessive overhead irrigation in the field or by lowering relative humidity in the glasshouse, reduces bacterial transmission and proliferation. Under wet conditions, farm workers and equipment may spread the disease from infected to healthy plants. Therefore, it is best to harvest when the foliage is dry. Fruit should be handled carefully to prevent wounds, for example by cutting rather than pulling the fruit off the vine. Copper-containing bactericides have been used to combat this pathogen, but often with limited success.

For cucumber, the disease is best controlled through the use of resistant cultivars. The original resistant type line is Gy 14 (*psl* gene), but cultivars such as 'Poinsett' slicer and 'Addis' pickle are resistant. Resistance in cucumber may be modified by several genes besides *psl*.

Astor yellows (Phytoplasma)

Astor yellows is caused by *Candidatus* Phytoplasma asteris (formerly a mycoplasmalike organism or MLO). Symptoms include intense yellowing, stunting of vines and excessive branching of affected plants. The disease affects summer squash with yellow fruit more than other cucurbits. The pathogen overwinters in weedy plants (dandelion, galinsoga, plantain, thistle and wild lettuce) and is transmitted to crop plants by the six-spotted leafhopper (*Macrosteles fascifrons*). Disease control is by eradicating the overwintering host plants from near crop production fields, as well as insecticides for the leafhopper.

Bacterial fruit blotch

Bacterial fruit blotch is a serious disease of watermelon. It first became important in 1989 in Florida and other southern states of the USA, and has been a problem there ever since. The causal agent, *Acidovorax avenae* ssp. *citrulli*, is seedborne and may be transmitted by means other than seed; for example, it is often mechanically spread in glasshouses where transplants are grown.

Infected seedlings develop dark water-soaked spots on the foliage and stems, becoming necrotic with yellow margins. Lesions also appear on the fruit, with cracking and exudation making the fruit unmarketable. High temperature and high humidity promote bacterial proliferation.

Disease assays of seed lots are recommended before planting that lot, and are done by growing seedlings in a glasshouse under ideal conditions for disease development. Testing helps to prevent a high incidence of infection, but cannot be relied on entirely to avert seedborne transmission. Seed lot testing is also done using polymerase chain reaction (PCR), a DNA amplification process, to test for seedborne bacterial fruit blotch.

Transplants should not be grown in the same glasshouse where there are plants from another seed lot contaminated with the bacterium. Copper-based foliar sprays help to control the disease. Planting in fields with recent crops of watermelon or other pathogen hosts, including citron, melon and eggplant, should be avoided. Resistance has been identified in the foliage of some accessions, as well as in the fruit of other accessions of *Citrullus*.

Bacterial leaf spot

Bacterial leaf spot of squash plants in India, Australia, Europe, the USA, Japan and elsewhere is caused by *Xanthomonas cucurbitae*. Cucumber and other cucurbits are sometimes attacked as well. The bacterium infects leaves through the stomata. Small (2–4 mm) brown lesions, surrounded by a yellow halo, develop on the leaves. Affected areas may coalesce, forming large dead regions delimited by veins. Diseased fruit will rot and may be a source of infection for fruit nearby. Overhead irrigation can spread the bacterium, which survives from one season to the next on plant material in the soil. The pathogen is also seedborne. Genetic resistance has been found in *Cucurbita moschata* and *C. okeechobeensis*.

Bacterial rind necrosis

Bacterial rind necrosis is caused by *Erwinia carnegieana*. Brown necrotic areas occur in the rind of melon and watermelon and sometimes extend into the flesh. Cultivars differ in degree of susceptibility. In watermelon, there may be some resistance in 'Jubilee' and 'Sweet Princess' compared with the more susceptible 'Klondike Blue Ribbon'.

Bacterial wilt

Bacterial wilt damages cucurbits in North America and, to a lesser extent, in Europe, Africa and Asia. It is seldom a serious threat to watermelon, but cucumber and melon plants may be killed. The bacterium (*Erwinia tracheiphila*) also infects squash, often resulting in an exudation and decay of the fruit from secondary soft rot organisms.

The first evidence of the disease is a dull green area on the leaf at the site of infection. The vascular system becomes plugged with a slimy substance, and one or two leaves, a branch, or the entire plant may suddenly wilt and die. Diseased plants are sometimes diagnosed by cutting a branch to determine if a sticky exudate will form strands to the knife when it is withdrawn from the cut surface.

E. tracheiphila overwinters in cucumber beetles. These insects disseminate the bacterium the following year when they begin feeding on cucurbits. Thus, controlling banded, striped and spotted cucumber beetles can reduce disease spread. Cucumber cultivars that are bitterfree (*bi-1* or *bi-2*), such as 'Marketmore 80' and most of the current greenhouse cultivars, are less preferred by cucumber beetles than other cultivars and thus have reduced infection. In cucumber, a single dominant gene (*Bw*) for resistance to bacterial wilt is known, but unfortunately is linked with the *m* gene for hermaphroditic sex expression (associated with round fruit).

Brown spot (internal fruit rot)

Brown spot, or internal fruit rot, of melon results from infection by *Pantoea ananatis* (= *Erwinia ananas*). Symptoms are yellow-brown spots up to 4 cm in diameter on the fruit. Damage by other organisms or by mechanical injury predisposes fruit to this disease, which is serious in a hot, humid environment.

Salmonella

Salmonella spp. can cause food poisoning. Cubed or sliced watermelon and melon, convenience foods for the consumer, are regulated under food safety laws in many countries. *Salmonella* and other species of bacteria on the external surface of the watermelon or melon can contaminate the flesh when the fruit is cut. Freshly cut watermelons and melons are wholesome and safe, but when prepared at the supermarket or restaurant salad bar, then bacteria can increase to dangerous levels if the cubes are held at warm temperatures.

Soft rot

Soft rot of cucurbit fruit, especially cucumber and melon, is caused by *Pectobacterium carotovorum* (= *Erwinia carotovora*) along with other organisms. Water-soaked blotches quickly develop into a wet, soft rot and the fruit collapses. This affliction is favoured by hot, humid conditions. Preventive measures include careful handling to avoid fruit injury, along with the use of chlorinated sprays or dips of the fruit. There are no resistant cultivars.

FUNGAL AND OOMYCETE DISEASES

Alternaria leaf blight

Alternaria leaf blight and storage fruit rot of winter squash, melon, cucumber, watermelon and other cucurbits are caused by *Alternaria cucumerina* and the newly described *A. citrullicola*, an important pathogen wherever cucurbits are grown. *Alternaria pluriseptata* and *A. alternata* also cause leaf spotting and fruit rot.

Light brown leaf spots, which are small and round at first, enlarge and cause defoliation. Concentric rings or a 'bullseye' appearance may develop at the lesions on the upper leaf surface. Yield and quality are adversely affected by the reduced photosynthetic area. Spores from diseased leaves infect fruit, especially if the fruit are predisposed by sunscald or chilling injury. Round sunken black or brown lesions, measuring about 1 cm in diameter, form and may cause the fruit to rot postharvest. Infection typically occurs at the peduncle scar or at a wound.

Either crop rotation, deep ploughing or removal of infected plant debris is recommended, since the fungus is capable of surviving on plant remains for 2 years. Overhead irrigation should be avoided in areas of high rainfall, because long periods of leaf wetness (> 8 h) encourage infection. Fungicides are effective and hot-water treatment after harvest can help to prevent fruit rot. Melon line MR 1 and cultivar 'Edisto 47' are resistant to the disease. Resistant watermelon cultivars include 'Sugar Baby', 'Fairfax' and 'Calhoun Gray'. A resistant cucumber cultivar is 'Marketmore 97'.

Anthracnose

Anthracnose can pose major difficulties for watermelon growers. The disease also attacks cucumber, melon, squash, bottle gourd, fluted pumpkin, pointed gourd, ivy gourd, chayote, snake gourd and other cucurbits. It is caused by *Colletotrichum orbiculare*, a destructive fungus found in India, southeastern USA and other warm, humid regions throughout the world. Two distinct races are

Fig. 9.2. Anthracnose of cucumber (*Cucumis sativus*), showing both leaf and stem lesions.

present in this pathogen: (i) race 1, which primarily infects cucumber and melon; and (ii) race 2, which primarily infects watermelon and melon. *Colletotrichum magna* also causes leaf and fruit spotting on cucurbits, but is less common.

Water-soaked areas on infected leaves turn into brown or black lesions and tan cankers (Fig. 9.2). Conidia from mature lesions are often disseminated by splashing rain. Stem lesions can girdle plants, causing wilting, and the plant may die. Sunken black circular or elongated spots occur on fruit, producing a gelatinous mass of pink or salmon-coloured spores when moisture is present. Old lesions become black and may have white centres with small black fruiting bodies. Lesions also develop on fruit in storage. The fungus does not require wounds on the fruit for infection.

Disease development is promoted by temperatures of 20–27°C (24°C is ideal) and 100% RH. The pathogen overwinters on infected vines and can be seed transmitted. Sanitation and rotation, disease-free seed, hot-water treatment of seeds, fungicides, low-temperature treatment of fruit after harvest and resistant cultivars are used to control anthracnose.

'Charleston Gray' was a leading watermelon cultivar in the USA for many years, mainly because of its resistance to anthracnose. 'Crimson Sweet' is another important cultivar with anthracnose resistance. The *Ar-1* allele, which is in the genetic background of a number of watermelon cultivars, provides resistance to races 1 and 3, but not to race 2 of anthracnose. Unfortunately, race 2 is the predominant race in most watermelon growing areas. 'AU-Producer', developed in Alabama, is resistant to race 2.

Anthracnose-resistant cucumber cultivars of the slicing type include 'Poinsett', 'Slicemaster' and 'Dasher II'. Many pickling cultivars, including 'Vlaspik' and 'Sumter', are also resistant.

Belly rot

Belly rot, due to *Rhizoctonia solani*, is a fruit rot of cucumber and melon. It is most serious in southern USA and other areas where warm, moist conditions prevail. Fruit in contact with damp soil may become infected and develop a water-soaked decay, often at that part of the fruit touching the soil. Infection spreads to other fruit during storage. The fungus also causes damping off, stem canker and root rot. It has a wide host range and is present in many soils. Removing infected plants and avoiding excessive irrigation may help to prevent disease spread. Resistance has been found in USDA plant introductions of cucumber.

Brown rot

Brown rot, also known as Choanephora wet rot, affects cucumber, melon, squash, luffa and other cucurbits in the tropics. Spores of the soilborne fungus (*Choanephora cucurbitarum*) are spread by the wind and insects and cause infection in rainy weather. Blossoms and young fruit become covered by fungal growth, which is white at first and later becomes purplish black. Young summer squashes and other cucurbit fruit may decay. This pathogen is able to survive in the soil for more than a year. Crop rotation and planting on well-drained soils aid in disease control. No genetic resistance is known in cucurbits.

Cercospora leaf spot

Cercospora leaf spot afflicts cucurbits worldwide, but most commonly is found on watermelon, melon and cucumber. The disease develops quickly when the temperature is above 25°C. The fungus (*Cercospora citrullina*) can cause defoliation, due to numerous lesions on the foliage, but is not known to infect fruit. Effective control measures include a good fungicide programme, sanitation by removing diseased vines and crop rotation. There are no resistant cultivars.

Charcoal rot

Charcoal rot, caused by *Macrophomina phaseolina*, attacks roots, foliage and fruit of many cucurbits. It is a minor disease that can be a problem under certain environmental conditions. The pathogen inhabits temperate and tropical areas worldwide, and is encouraged by high temperature and soil moisture. Plant parts in contact with the soil develop lesions with black microsclerotia. Cankers, often with concentric rings, occur on the stem and may girdle the plant. Excessive overhead irrigation, and other conditions favouring free moisture on the fruit and foliage, should be avoided. Soil fumigation has met with limited success. There are no resistant cultivars.

Cottony leak (oomycete pathogen)

Cottony leak, caused by *Pythium aphanidermatum* and other species of *Pythium*, is a fruit rot of cucumber, squash, melon and watermelon. Cottony fungal growth develops from water-soaked lesions at infection sites and spreads over the fruit, which may produce a watery discharge. These pathogens also cause damping off of seedlings and a root rot.

Pythium species have a very wide host range and persist in soil for years, making rotation largely ineffective for control. Moist conditions and warm temperature in the field or during storage promote disease development. Plastic mulch should be avoided in rainy areas, because the depression where the fruit sits will collect water. However, the benefits of using plastic outweigh the risk for many growers. Artificial barriers such as wood blocks can be placed between the fruit and soil to avoid infection.

Cottony soft rot

Cottony soft rot, due to *Sclerotinia sclerotiorum*, produces stem cankers and fruit rots on many cucurbit crops worldwide. The first sign of this disease is usually white, cottony fungal growth on the stem. Infected plants may develop a white mould with black sclerotia inside the stem and then die. The fungus also grows on leaves and fruit. The pathogen is a necrotroph and ascospores require dead tissue to germinate and grow, so fruit are often infected through a withered flower still attached to the developing ovary. A soft rot and white cottony mould with embedded black sclerotia quickly envelops the fruit. The disease spreads by contact from one fruit to another in storage.

The fungus also causes white mould of beans, cabbage and more than 400 other plant species. Build-up of this pathogen in a field planted with one of these crops may lead to considerable destruction of squash or melon plants grown in that field the following year. Since there are other pathogen hosts and because sclerotia survive in the soil for years, rotation away from cucurbit crops may not provide adequate control. Sanitation, fungicide application, care in irrigating and harvesting, good storage conditions (10–15°C and 50–75% RH) and the prompt removal of infected fruit from storage are recommended control measures.

Damping off (oomycete pathogen)

Damping off, a problem of worldwide distribution, can be incited by several pathogens, including *Pythium* spp., *Phytophthora* spp., *Rhizoctonia solani* and other fungi. Damping off often causes poor germination and death of seedlings soon after emergence. It is particularly serious in cool, damp weather, which delays germination and extends the time during which seedlings are most vulnerable to infection. The pathogens responsible for damping off of seedlings can sometimes also cause a root rot in older plants. Stems and roots of infected cucurbits become water-soaked and flaccid, and the foliage may wilt and die if the infection is severe enough (Fig. 9.3). If disease severity is low, cucurbits can outgrow the disease. However, if plants survive to maturity, yield may still be reduced.

Seed treatment with a broad-spectrum thiram-based fungicide reduces damping off, and treated seeds are often available. Disinfestation of plant containers and growing media can eliminate some sources of inoculum. Shallow planting, delaying seeding until warm weather, raised beds and other procedures that speed germination may also lower the incidence or severity of the disease. Cultural practices to keep foliage dry and minimize standing water, such as wide plant spacing, raised beds and drip irrigation, can reduce damping off. When starting plants in the greenhouse, avoiding prolonged overhead irrigation can be helpful. Cucurbit seeds should not be sown in poorly drained soils. Rotation, by not planting in fields where cucurbits or other hosts of the pathogens were grown the previous year, should be practised.

Downy mildew (oomycete pathogen)

Downy mildew, produced by the oomycete *Pseudoperonospora cubensis*, can be recognized on the leaves by the yellow angular lesions delimited by veins on the upper leaf surface and grey/brown sporulation below (Fig. 9.4). This disease, which leads to defoliation and greatly reduced yield, occurs in eastern USA, Mexico, Israel, Europe, Asia and many other areas worldwide. Although cucumber and melon are particularly affected, squash, watermelon, luffa, pointed gourd, ivy gourd and many other cucurbits are susceptible as well. This pathogen only infects members of the Cucurbitaceae.

Fig. 9.3. Phytophthora crown rot on zucchini seedling. Water-soaked lesion and cracking, with white sporangia starting to form on the lower section of the lesion.

Fig. 9.4. Downy mildew on cucumber leaves. (a) Chlorosis and necrosis on the abaxial side of the leaf. (b) Grey sporulation on the adaxial side of the leaf.

Downy mildew is favoured by moist conditions. It is an important field disease and can be serious in glasshouses with high humidity. The pathogen overwinters on cucurbits in tropical and subtropical areas; windblown spores from there can later infect plants in temperate areas. It may also persist as oospores on plant debris.

Prior to 2004, resistant cultivars were widely used to control the disease in cucumber. Since 2004, the pathogen has overcome genetic resistance and fungicides are the only effective method employed to combat downy mildew. Despite chemical rotations, this pathogen rapidly acquires resistance to fungicides, and in the eastern USA this is the primary pathogen limiting production. Cultural practices have had little effect without the addition of a robust fungicide regime. New sources of durable resistance have been identified in wild PI accessions and cucumber breeding programmes worldwide are working to incorporate downy mildew resistance into commercial cultivars, the first of which are now available (e.g. Bristol slicer).

Fusarium fruit rot

Fusarium fruit rot, which can be caused by *Fusarium solani* f. sp. *cucurbitae*, is an important disease of cucurbits, particularly squash. *F. solani* f. sp. *cucurbitae* and the related species *F. petroliphilum* attack roots and stems, while *F. solani* f. sp. *cucurbitae* is the primary pathogen on fruit of the host plant. The pathogen invades the outer part of the stem, causing wilting, brown decay at the base of the stem and collapse of the plant. The stem may become covered with the white or pink fungal growth. Infected fruit have water-soaked areas that develop into a brown or grey dry rot. Other *Fusarium* species, including *F. equiseti*, cause cucurbit fruit rots in which infected areas on stored fruit turn pink, red or purple, with white or pink mycelia.

Because spores are carried externally on cucurbit seeds, seed treatment is used to reduce disease spread. The fungus is persistent in the soil, and a rotation of at least 3 years is recommended. Hot-water treatment (1 min at 57°C) and fungicide application will limit pathogen transmission among harvested fruit. There are no resistant cultivars.

Fusarium wilt

Fusarium wilt, caused by *Fusarium oxysporum*, is an important disease of melon and watermelon, especially in the Americas, Europe and Asia. It also attacks cucumber. The soilborne fungus penetrates the roots and develops in the vascular system, interfering with water movement and causing a severe wilt. It flourishes when soil temperature is high. Watermelon is particularly affected and diseased seedlings may die.

Preventive measures include rotation with crops that are not hosts of the pathogen and not planting susceptible cucurbits in fields with a known history of Fusarium wilt. Because the fungus exists in the soil for many years, short rotations are inadequate. Watermelon should be rotated with crops such as corn, cotton and legumes with a cycle of 5–7 years.

Raising soil pH to 6.5 inhibits disease development. In Asia, Fusarium wilt has been controlled by grafting melon plants on to Malabar gourd, wax gourd, bottle gourd, or interspecific hybrid squash rootstock. Although hot-water seed treatment reduces seed transmission, the use of disease-free seeds, produced where the pathogen does not occur, is recommended.

Different biotypes of the fungus cause the Fusarium wilts of different cucurbits. Fusarium wilt of watermelon is caused by *Fusarium oxysporum* f. sp. *niveum*, Fusarium wilt of melon is due to *F. oxysporum* f. sp. *melonis* and that of cucumber is caused by *F. oxysporum* f. sp. *cucumerinum*. Three races of watermelon Fusarium wilt have been identified. Resistant watermelon cultivars have been available since the first decade of the 20th century; 'Charleston Gray', 'Summit' and other cultivars are resistant to race 0, and newer hybrids are resistant to race 1. Four races of Fusarium wilt of melon are known. Melon cultivars resistant to races 0 and 2 include 'Perlita', 'Infinite Gold', 'TAM Dew Improved' and others; 'Charmel' is resistant to race 1, while 'Athena' is resistant to races 0, 1 and 2. USDA release 'MR-1' is resistant to all known races. Distinct races of Fusarium wilt of cucumber have also been discovered. In cucumber, the genes for resistance to Fusarium wilt and scab are linked. 'Poinsett' slicer and 'Wisconsin SMR 18' pickle are resistant.

Grey mould

Grey mould, due to *Botrytis cineraria*, is a problem for cucumber plants grown in cool, cloudy weather. It attacks glasshouse cucumbers everywhere and also afflicts melon and squash. Infection often occurs through dead tissue, such as senescent petals and other flower parts. Abundant mycelia and spores develop on the blossom end of the fruit, causing it to rot. Brown lesions form on the foliage and are covered by a mass of grey spores. Good ventilation, warm temperature and avoiding free moisture on the leaf surface are preventive measures in the glasshouse.

Gummy stem blight and black rot

Gummy stem blight and black rot are caused by three species of *Stagonosporopsis*: *S. cucurbitacearum* (syn. *Didymella bryoniae*), *S. citrulli* and *S. caricae*. The pathogen produces a brown, blotchy blight with black spots on foliage of watermelon, cucumber, melon, squash, bitter gourd, wax gourd, chayote and other cucurbits when conditions are hot and humid. It is an important field disease in southern USA, South America and other subtropical and tropical growing areas, and can be a serious concern for glasshouse cucumber growers. Black rot is also a market and storage disease of cucurbit fruit.

Infection in the seedling stage is often lethal. Tan to black lesions form on the foliage and fruit of older plants. The disease may develop so suddenly that foliage appears to be scorched. The stem can become infected, girdling and killing the plant. Black fruiting bodies may form on foliage and fruit of some cucurbits.

Infected *Cucurbita pepo* and *C. maxima* fruit typically turn black and quickly rot, but 'Butternut' (*C. moschata*) fruit develop circular brown rings without a soft rot. 'Butternut' squashes often display fruit symptoms in the field, but those of *C. maxima* usually do not succumb to black rot until after harvest. Fruit may become infected through wounds, and careful handling during harvest and shipping can minimize losses. Curing harvested fruit (see Chapter 7 for more details) and low postharvest temperature (but > 10°C) reduce the incidence of black rot during storage.

Infected seeds spread the fungus. Consequently, only disease-free seeds should be planted. Removing afflicted plants and crop rotation for 2 years or more are advised. Ventilation in the glasshouse and other measures to reduce humidity are helpful. To avoid infection, irrigation methods that do not leave moisture on leaf surfaces are preferred. Fungicides should be applied early in the day so that the plants do not remain wet overnight. Fungal resistance to benzimidazole and other fungicides has evolved in some glasshouse growing areas in Europe and in cucurbit fields in eastern USA.

Limited genetic resistance has been found in cucumber, melon, squash, watermelon and wild species of *Cucurbita* and *Cucumis*. 'Congo' watermelon, 'Nicklow Delight' squash (*C. moschata*) and 'Gulfcoast' melon possess resistance alleles.

Monosporascus root rot and vine decline

Root rot and vine decline of cucurbits result from infection by a number of organisms (Blancard *et al.*, 1994). One of these pathogens, *Monosporascus cannon-ballus*, creates a serious disease problem for melon and watermelon plants grown in Texas, Arizona, California, Tunisia, Israel, Japan and Spain. Also affected are cucumber, squash, smooth luffa and bottle gourd. The pathogen is adapted to hot, arid climates. The fungus infects the root system of young plants, killing most of the feeder roots and producing black perithecia on the lateral roots. It causes yellowing and necrosis of crown leaves, often followed by a sudden collapse of the entire plant shortly before maturity. The canopy may be lost 5–10 days after symptoms first appear. Control is by long-term rotation with non-susceptible crops. Melon cultivars are being bred for resistance: 'Honey Dew' and 'Deltex' are tolerant. *Cucurbita* rootstocks have also been used to manage this disease in high-risk areas, where large profit margins can offset the expense of grafting.

Net spot

Net spot is a disease of cucumber that also affects balsam pear (*Momordica charantia*). The pathogen is *Leandrea momordicae*. It causes white leaf spots with reddish margins on the upper and lower leaf surfaces and causes severe defoliation of cucumber plants. No control methods are known.

Phomopsis black root rot of cucumber

Black root rot, incited by *Diaporthe sclerotioides*, occurs in northern Europe, Asia and Canada. It is primarily a malady of glasshouse-grown cucumber and melon, but affects other cucurbits as well. The soilborne fungus invades root hairs, causing a root rot. Root symptoms are sunken, irregular, grey-ish-black spots surrounded by a darker line. Vines may wilt near harvest time, depending on the degree of root infection. It can survive in the soil for several years and thus is difficult to control by rotation. Sanitation helps to prevent dissemination via contaminated soil and the disease can be avoided by the use of soilless growing media. Systemic fungicides and soil fumigation are effective. Melon and cucumber plants are sometimes grafted to resistant Malabar gourd (*Cucurbita ficifolia*) rootstock.

Phytophthora root and crown rots (oomycete pathogen)

Phytophthora root rot and crown rot, due to *Phytophthora capsici* and other species of *Phytophthora*, cause plants to wilt rapidly and die, often during the

Fig. 9.5. Spaghetti squash displaying symptoms and signs of *Phytophthora capsici* fruit rot.

fruiting stage. This soilborne oomycete infects many cucurbits as well as other crops, causing foliage blight, with black lesions developing at the crown. Roots and crowns of diseased plants have a water-soaked appearance and a black or brown decay. *P. capsici* also produces a fruit rot of melon, cucumber, squash and watermelon (Fig. 9.5). Soft, sunken spots develop on the fruit and a white mould may form there.

Disease development is promoted by warm weather and wet soil. Good drainage helps to prevent Phytophthora root rot. Damage can be averted by proper cultural practices, such as the use of raised beds and drainage ditches, avoiding overly prolonged irrigation or using surface water, and not compacting wet soil with heavy machinery. Preventing fruit contact with soil can reduce disease. The pathogen can survive in the soil for more than 15 years and has a broad host range, making crop rotation inadequate. Systemic fungicides are also an effective means of disease control. Sources of resistance to Phytophthora root and fruit rot have been identified in cucumber, watermelon and squash, so resistant cultivars should become available in the future. Ontogenic or age-related resistance is common, with even susceptible cultivars showing reduced disease with fruit age.

Powdery mildew

Powdery mildew is an annual problem for cucurbits and other crops everywhere. *Podosphaera xanthii* (syn. *Sphaerotheca fuliginea* and *Podosphaera fuliginea*), *Leveillula taurica* (anamorph = *Oidiopsis taurica*) and *Golovinomyces cucurbitacearum* (formerly *G. cichoracearum* and *Erysiphe cichoracearum*) cause powdery mildew. It is difficult to distinguish among the species because the most diagnostic feature is the cleistothecium, which rarely occurs on cucurbits. Scientists previously thought that *G. cucurbitacearum* was the causal agent of powdery mildew in most areas, but now *P. xanthii* is considered the primary agent in Europe and the USA. *G. cucurbitacearum* has a lower optimal temperature than *P. xanthii*, thus the latter is more widespread during the warmer summer and autumn months.

First signs of infection are circular white powdery spots, which later enlarge and coalesce until the entire leaf is covered (Fig. 9.6). The white colonies inhabit the upper and lower leaf surfaces. The leaves turn yellow, then brown, and die prematurely. This decrease in photosynthetic capability results in reduced yield and fruit quality, particularly in melon.

Powdery mildew fungi are unique, in that the mycelia or hyphae only grow on the surface of the leaf. The haustoria, responsible for nutrient uptake and communication with the host, is the only part of the fungus to reside inside the plant cells. External mycelia and sporulation make these fungi vulnerable to environmental factors. For example, heavy rains can wash away some spores and mycelia, while intense sunlight can damage spores. Consequently, disease development is favoured by warm temperatures ($\geq 15°C \leq 30°C$) and high relative humidity (> 94%).

Sulfur is a classic but effective means of controlling powdery mildew. Wettable or micro-encapsulated sulfur is applied to cucurbit foliage in the field, while in glasshouses sulfur pellets are vaporized. At high temperature and full sun, sulfur can be phytotoxic. However, some melon cultivars, including 'V-1' and 'SR 91', are able to tolerate sulfur even under hot conditions.

Protective fungicides, which tend to operate with multiple modes of action, help to control powdery mildew when all plant surfaces can be reached. For systemic fungicides, not all plant surfaces must be contacted with the pesticide to achieve control. Fungicide resistance has historically developed faster in systemic fungicides, which tend to rely on one or two modes of action to control the pathogen. Fungicide resistance occurs in *P. xanthii* to benomyl, triadimefon and at least four other active ingredients used to combat the disease (Zitter *et al.*, 1996).

Powdery mildew resistance in cucumber is governed by several recessive alleles. Resistant cultivars include 'Marketmore 76', 'Polaris 76', 'Pixie' and others. There is a genetic association in cucumber between resistance to

Fig. 9.6. White powdery mildew colonies covering a cucumber leaf. (Photo: Dr Lina Quesada, NC State University.)

powdery mildew and to downy mildew. Although the pathogens causing these two diseases are quite distinct, one being a fungus and the other an oomycete, cucumber cultivars resistant to one mildew disease are generally resistant to the other. Genetically associated with resistance to these diseases is a physiological chlorosis and necrosis that develops in cucumber under certain environmental conditions.

Powdery mildew resistance in melon is race specific. 'PMR 45' has a single dominant allele for resistance. It was a leading cultivar in western USA and other areas for more than 50 years, but is only resistant to race 1 of powdery mildew. 'PMR 5', 'PMR 6' and some other cultivars are resistant to races 1 and 2 as a result of a single dominant allele and modifiers. Breeding line MR-1 is resistant to all three races.

In squash and other *Cucurbita* species, cultivars with tolerance or intermediate resistance are available for butternut and acorn squashes, pumpkins, yellow summer squash and zucchini. In *C. pepo*, those cultivars derived from the Mississippi River Valley centre of domestication show greater tolerance to this disease compared with those from the Mexican centre of domestication (Cohen *et al.*, 2003). Resistance found in wild *C. okeechobeensis* ssp. *martinezii* has been transferred via bridge crossing to *C. moschata* into *C. pepo* squash cultivars (Cohen *et al.*, 2003). This resistance, designated *Pm-0*, behaves as a single partially dominant gene and is the main form of resistance used in contemporary *C. moschata* and *C. pepo* commercial cultivars (Jahn *et al.*, 2002; Holdsworth *et al.*, 2016). Two partial resistance genes (*pm-1* with three alleles and *pm-2*) have been reported in *C. moschata* but do not appear to have been deployed beyond this species. Powdery mildew resistance has been observed in *C. maxima* and *C. ecuadorensis*, but its genetic basis has not been studied.

This disease attacks watermelon in some areas of southwestern Asia, but it is not a problem for this crop in most countries. Seshadri (1986) reported that powdery mildew affects watermelon plants in southern India, but not in northern India, presumably due to biotype differences. 'Arka Manik' is resistant to the powdery mildew race infecting other watermelon cultivars in Bangalore.

Powdery mildew resistance in bitter gourd is governed by many recessive alleles. Bitter gourd breeding lines resistant to powdery mildew have been developed at the World Vegetable Center.

Powdery mildew caused by *Leveillula taurica* attacks cucumber and other cucurbits in the Mediterranean region and has a wide range of host species in other plant families. This fungus is controlled by the same fungicides used against other powdery mildew pathogens of cucurbits.

Rhizopus soft rot

Rhizopus soft rot, caused by *Rhizopus stolonifer* and related species, produces a postharvest fruit rot of melon, squash, cucumber, watermelon and other cucurbits. The disease is promoted by the presence of free water and by fruit cracks or wounds, which may become sites of infection. Infected tissue becomes soft and water-soaked. Careful handling and the maintenance of low temperature during shipping and storage discourage disease development. These fungi have a very wide host range and survive in the soil from one year to the next.

Scab

Scab is primarily an affliction of cucumber plants grown in northern areas of the USA, Europe and Asia, where it is cool and moist. The causal agent (*Cladosporium cucumerinum*) also attacks melon, watermelon and squash.

Water-soaked spots form on the leaves of infected plants, later turning white with a chlorotic halo around the lesions. The fungus kills the apical meristem and diseased seedlings may die. Sunken dark blotches, sometimes with a sticky exudate, occur on fruit, making them unmarketable. Corky scar tissue develops at the edges of lesions on melon and squash fruit. Melons also become infected at the peduncle scar, in which case a fruit rot ensues.

The fungus overwinters on dead vines of diseased plants and it can also be seedborne. Scab is controlled by protectant fungicides, but multiple applications may be required, especially during long periods of cool wet weather. Other control measures include the use of rotation and cultural practices that minimize the presence of free moisture on the foliage.

A single dominant allele for resistance in cucumber has been found and bred into many cultivars, both slicers and picklers. Despite the concern of some for the vulnerability of single gene resistance, scab resistance in cucumber has proved to be stable and durable. Scab-resistant cultivars have been grown for nearly 50 years, without resistance breaking down due to the occurrence of new fungal races.

Septoria leaf spot

Septoria leaf spot is a leaf blight and fruit spot of melon and squash caused by *Septoria cucurbitacearum*. The pathogen thrives in temperate areas with rainy weather and temperatures of 16–19°C. Small white spots with brown halos appear on leaves of infected plants. White spots may also occur on the fruit, making them unmarketable. Pycnidia, which are small black fruiting bodies, develop in the lesions. The fungus overwinters on infected plant debris; hence, rotation and ploughing after harvesting an infected field can help to prevent persistence of the disease. The pathogen is generally controlled by chemical means, even though some fungal strains have developed resistance to some fungicides.

Southern blight

Southern blight is caused by a related soilborne fungus (*Athelia rolfsii*), which produces a fruit rot of melon, squash and watermelon, especially in tropical and subtropical areas. Infected plants often wilt and turn yellow. They may die due to girdling of the stem at its base. White fungal growth with embedded brown sclerotia develops on the stem and on fruit in contact with the ground. Rotation, sanitation and raising soil pH to 7.0 aid in the containment of this disease. Resistant cultivars are not available and fungicides are not effective.

Stem end rot

Stem end rot, a product of *Lasiodiplodia theobromae*, is a common storage disease of watermelon in the Americas. It also afflicts melon, squash, cucumber and other vine crops.

Although the disease is primarily a postharvest problem, causing rot of fruit in shipment, in storage or at market, the fungus also proliferates in the field. It causes wilting and a foliage blight, producing lesions with a gummy exudate on stems and leaves. Infected fruit tissue becomes soft and water-soaked; later, the watermelon shrivels, turning brown at the peduncle end. Dark grey mycelia and black fruiting bodies form in the lesions and secondary organisms may grow there as well. Diseased fruit sometimes have an odour.

Careful handling to prevent fruit injury helps to avert infection. Watermelons should be cut, not pulled, from the vine at harvest, and a long peduncle should be left on the fruit. For shipping, the peduncle can be recut and treated with a copper-containing paste. Low temperature retards development of the fungus and appropriate postharvest conditions reduce losses from stem end rot. Application of fungicide is important for control in the field.

Sudden wilt (oomycete pathogen)

Sudden wilt of melon is an abrupt collapse of the plant, often just before harvest, when plants are stressed by a heavy fruit load. It frequently occurs following cold nights. Various organisms, which might not be the cause, have been isolated from affected plants, but *Pythium ultimatum* and related species and CMV have been implicated as causal agents. Sudden wilt can be overcome completely by grafting melon to interspecific hybrid rootstocks.

Target leaf spot

Target leaf spot of cucumber and other cucurbits, which is due to *Corynespora cassiicola*, has occurred for many years in European glasshouses. It is also a problem for field crops in the USA and many other temperate, subtropical and tropical countries.

Target leaf spot has been called the 'leaf fire' disease in reference to the yellow lesions and necrotic spots that develop on foliage of afflicted plants. Angular lesions turn grey and later drop out, giving a shredded appearance and causing defoliation. Young fruit that become infected at the blossom end will shrivel.

The fungus can survive for 2 years on infested plant material and conidia are airborne. Disease development is favoured by long hot humid days, especially when nights are considerably cooler.

Infected vines, a source of inoculum, should be removed from the field and fungicide applications are recommended. Resistance in cucumber is associated with susceptibility to powdery mildew. 'Marketmore 97' cucumber is resistant to target leaf spot.

Ulocladium leaf spot

Ulocladium leaf spot, due to *Ulocladium cucurbitae*, produces a leaf blight on field-grown cucumber plants in New York and on field- and glasshouse-grown cucumbers in the UK. Irregular tan lesions with a buff centre, measuring 3–30 mm in diameter, develop on leaves and coalesce to cause defoliation. This minor disease can be controlled by fungicides, rotation or the use of resistant cultivars such as 'Marketmore 97'. *Ulocladium atrum* also occurs on cucumber in the UK and France, but is usually of little importance.

Verticillium wilt

Verticillium wilt, due to *Verticillium dahliae* or *V. albo-atrum*, attacks melon, squash, cucumber, watermelon and other cucurbits worldwide. It is a soilborne field disease that also damages glasshouse cucumbers. Symptoms are similar to those of Fusarium wilt, but Verticillium wilt can be serious in areas with lower soil temperature. Wilting often occurs near plant maturity, and the chlorosis and necrosis of crown leaves may be more severe than with Fusarium wilt.

V. dahliae causes cavity rot of wax gourd fruit. Stored fruit may appear outwardly sound, but internally they contain decay and mycelia. The fungus also invades wax gourd roots, stems and peduncles.

Verticillium wilt fungi can persist in the soil for years, so cucurbits should not be replanted in a field with a history of the disease. Melon cultivars differ in their susceptibility to Verticillium wilt. 'Persian' and 'Casaba' are more easily infected than 'Honey Dew' and 'PMR 45'.

NEMATODES

Root knot nematodes (*Meloidogyne incognita* and other *Meloidogyne* species) invade root tissue, causing galling of cucurbit roots. Adult nematodes are embedded in the root galls. Infected plants are stunted and wilted and yield is reduced. Nematodes also damage cucurbits by transmitting tobacco ringspot virus (TRSV).

Root knot nematodes cause substantial losses of melon, watermelon, squash, cucumber and many other cucurbits. They are usually most prevalent on sandy soils and are seldom serious in the fields of northern areas, where severe winter temperatures inhibit their overwintering. However, they may infest soil beds in glasshouses anywhere.

Control measures include soil fumigation, rotation with a crop not included in the wide host range of the nematodes affecting cucurbits, and grafting to a resistant rootstock. Cucurbits that are resistant include African horned melon, West Indian gherkin, wax gourd and citron. African horned melon has been used as a resistant rootstock, but attempts to transfer its resistance to melon and cucumber have been unsuccessful. Resistance to some species of *Meloidogyne* was found in *Cucumis sativus* var. *hardwickii*, but this taxon was susceptible to both races of *M. incognita* tested (Walters *et al.*, 1993). Cucumber, watermelon and melon cultivars differ in their susceptibility to species of *Meloidogyne*, but none is resistant.

An experiment with cucumber plants grown on nematode-infested soil (Hanna *et al.*, 1993) tested the effects of double-cropping with a nematode-resistant cultivar of tomato, using cucumber transplants instead of direct seeding, and pre-plant field application of the nematicide ethoprop. Preceding cucumbers with a nematode-resistant tomato significantly reduced nematode populations in the field. Although using cucumber transplants and applying ethoprop did not reduce overall nematode population size, cucumber yield and growth were improved.

VIRAL DISEASES

Over 30 viruses infect cucurbits, 11 of which are seedborne (Keinath *et al.*, 2017). Viruses can be a very serious problem, because there are no chemical protectants against them. The best control measure is a resistant cultivar, if available. Eliminating weeds and controlling whiteflies and other insects that transmit the virus can also be beneficial. However, aphids may carry a virus to cucurbits from far away, so weed control alone in a cucurbit field may not contain the disease. There has been some success with reflective mulch to repel aphids, and rowcovers can exclude insect vectors altogether.

Beet pseudo-yellows virus

Beet pseudo-yellows virus (BPYV), previously muskmelon yellows virus, infects melon and cucumber plants in Spain, The Netherlands, Japan and other countries. It has become especially troublesome on melon plants grown in plastic tunnels and other protected environments. BPYV causes chlorosis and reduces yield and quality. Similar symptoms are produced on cucumber and squash in The Netherlands and Japan by pathogens considered to be other strains of BPYV. The virus overwinters on a wide range of weed hosts and is carried from these hosts to cucurbits by the glasshouse whitefly. Chemical treatments to control whiteflies may restrict spread of the disease, but seldom can a large whitefly population be eradicated and virus transmission eliminated. Disease spread is reduced by sanitation, weed control and rowcovers that exclude insects. Tolerance has been found in Asiatic melon landraces. African horned melon and teasel gourd are resistant.

Clover yellow vein virus

Clover yellow vein virus (CYVV) was formerly known as the severe strain of bean yellow mosaic virus. It produces small, quite noticeable spots on leaves of infected squash plants. CYVV can cause economic losses to cucumber, but is usually not as destructive to squash.

Cucumber green mottle mosaic virus

Cucumber green mottle mosaic virus (CGMMV) attacks cucumber, melon and watermelon in subtropical areas, including India. It is also a serious affliction of glasshouse cucumbers in Europe, Japan and elsewhere. Infection occurs through the roots, coming from diseased plant debris in the soil. Leaf distortion, mottling and vein clearing occur and yield may be reduced. Symptoms vary according to the strain of the virus.

The virus is seed transmitted; hence, the use of disease-free seeds is important. Seed treatment with high temperature and trisodium phosphate (Na_3PO_4), or another appropriate chemical, is recommended. Sanitation is vital, since the pathogen can be transmitted from roots of previous crops and is also spread mechanically. Resistance to CGMMV has been found in melon.

Cucumber mosaic virus

Cucumber mosaic virus (CMV) stunts seedling growth (Fig. 9.7), causes mottling and distorted shape of leaves and fruit, and reduces yield. Cucumber, squash, melon and many other cucurbits are susceptible, but watermelon is less affected by most strains of CMV. The disease is of worldwide distribution and importance.

CMV is spread by aphids from its very wide range of host plants. Cucumber beetles also carry the disease. Insecticides used to kill the vectors are often ineffective for control of CMV, since the insects may disseminate the virus before being killed. The virus is easily transmitted mechanically.

CMV resistance seems to be common among wild *Cucurbita* species. Interestingly, all wild taxa tested by Provvidenti *et al.* (1978a,b) were resistant to this important virus except *C. pepo* var. *texana* and *C. maxima* ssp. *andreana*, the closest wild relatives of two of the domesticated squash species. Resistance has been found in some landraces of *C. maxima* and *C. moschata* from South America and Nigeria, respectively. A single dominant gene (*Cmv*) has been described for resistance from *C. moschata* 'Nigerian Local' (Brown *et al.*, 2003).

Most modern cucumber cultivars are resistant to CMV. In contrast, all important melon cultivars are susceptible, but resistance occurs in some members

Fig. 9.7. Squash (*Cucurbita pepo*) seedlings inoculated with cucumber mosaic virus. The seedling on the left was resistant; the one on the right was susceptible and shows symptoms.

of the Conomon Group. In recent years, genetic engineering techniques have been employed to transfer CMV coat protein-derived resistance to cultivars of squash, cucumber and melon; see Chapter 3 for more details.

Cucurbit yellow stunting disorder virus

Cucurbit yellow stunting disorder virus (CYSDV) produces symptoms similar to lettuce infectious yellow virus (LIYV), though it is more widespread and has become a serious threat to melon production in subtropical regions worldwide. It was first detected in the USA in south Texas during the spring of 2000. It is now one of the most common and damaging melon viruses throughout Latin America, as well as Texas, Arizona and California. Suppression of the silverleaf whitefly vector has been the most effective means of control. However, insecticide-resistant populations of this pest have arisen and continue to spread the virus. Recessive resistance from an African wild melon, 'TGR 1551', holds some promise for developing resistant cantaloupe and honeydew cultivars.

Melon necrotic spot virus

Melon necrotic spot virus (MNSV) produces necrotic lesions on the foliage of melon and cucumber in Mediterranean countries, Japan and elsewhere, especially on crops grown in glasshouses or other protected environments. The disease is transmitted by a soil fungus, *Olpidium bornavanus*, which enables the virus to enter the plant through the roots. This fungus also transmits cucumber necrosis virus and tobacco necrosis virus to cucumber. The virus is readily spread mechanically and can be seedborne as well. Consequently, only disease-free seeds should be planted. Sanitation, rotation, soil fumigation and grafting to a resistant rootstock of Malabar gourd are preventive measures.

Melon virus

Ourmia melon virus (OuMV) is the type species of a novel virus genus, so far restricted to melons in Iran. It causes mosaic, yellowing and chlorotic spots on leaves of infected plants. Mechanical transmission has been demonstrated, but no insect or mite vector has yet been identified.

Papaya ringspot virus

Papaya ringspot virus (PRSV) type W, formerly known as watermelon mosaic virus-1, is a serious disease of many cucurbits in warm regions worldwide. It causes leaf mottling and distortion, and infected fruit may become bumpy and deformed. The virus overwinters on cucurbits in tropical and subtropical areas and can be dispersed from there to other areas by aphids. Sanitation and cultural procedures to control aphids restrict dissemination of the disease. 'Sweet Slice', 'Douceur' and 'Zena' cucumbers, 'Redlands Trailblazer' squash and 'Cow Leg' bottle gourd are resistant.

Within squash, *C. pepo* is completely susceptible, but resistance has been found in *C. ecuadorensis* and *C. moschata* (Provvidenti, 1978a,b; Brown *et al.*, 2003). A single recessive gene (*prv*) was described within *C. moschata* (Brown *et al.*, 2003). However, when transferred to *C. pepo*, genetic control becomes more complex and it is difficult to attain levels of resistance similar to that found in the *C. moschata* background. Resistance found in *C. ecuadorensis* has been transferred into *C. maxima* squash cultivars (Herrington *et al.*, 2001).

Squash leaf curl virus

Squash leaf curl virus (SLCV) has become a problem in recent years for squash growers in southwestern USA and Mexico. Related viruses attack watermelon and melon in the Middle East and elsewhere. SLCV causes a severe leaf curl (with margins curling upward), mottling and enation. Infected fruit may be small, deformed and unmarketable.

The virus, which is prevalent in subtropical areas where cucurbits are continuously cropped, is transmitted by the sweet potato whitefly. Avoiding growing cucurbits and other hosts of this insect for a period of time in an area can keep the disease in check by curtailing build-up of the vector. There are no resistant cultivars, although tolerance has been reported in *C. pepo*, *C. moschata* and wild species of *Cucurbita* and *Cucumis.*

Squash mosaic virus

Squash mosaic virus (SqMV), which has a worldwide distribution, is a very destructive disease of squash and melon. The virus causes severe leaf distortion, often resembling injury from auxin-type herbicides. There may be enation on the lower leaf surface. Fruit are frequently mottled and deformed, and yield reduced.

Insects, including cucumber beetles and *Epilachna chrysomelina*, are contagion vectors. SqMV overwinters in cucumber beetles and infected seeds. To prevent seedborne infection, only virus-free seeds should be planted and seeds should not be saved from afflicted plants. Insect control and removal of diseased plants reduces spread of the virus. Weed control is helpful, since *Chenopodium* species can become infected with SqMV and transmit the pathogen by seed. Immunity or high-level resistance to SqMV has not been found in *Cucurbita*, but Provvidenti *et al.* (1978a,b) reported that *C. ecuadorensis* recovered from symptoms after infection. Resistance exists in bottle gourd.

Tobacco ringspot virus

Tobacco ringspot virus (TRSV) is transmitted to squash, cucumber, watermelon and other cucurbits by nematodes. It can also be spread mechanically and by insects. Tiny yellow-brown spots with a yellow margin occur on leaves. Small water-soaked spots may develop on fruit. The virus has a very wide range, overwintering on many weed species of different plant families. Weed control can help to keep the virus in check. Soil fumigation with nematicides curtails transmission by nematodes, but at an expensive cost. Resistance has been found in melon, squash (Provvidenti *et al.*, 1978a,b) and bottle gourd.

Tomato leaf curl New Delhi virus

Tomato leaf curl New Delhi virus (ToLCNDV), is a recombinant member of the bipartite begomovirus complex, which has moved into cucurbit crops. Vectored by whiteflies, it has spread from Asia to the Mediterranean basin and poses a serious threat to melons and cucurbits in Spain and other countries. Symptoms include leaf curling, vein swelling, severe chlorosis, fruit roughness and plant stunting. Some sources of resistance in *Cucurbita moschata* have been identified and efforts to breed resistant squash cultivars have begun.

Tomato ringspot virus

Tomato ringspot virus (TRSV) is spread to cucurbits by dagger nematodes. Squash is the most affected by this virus with symptoms including reduced leaf size, short internodes, and ring spots on discoloured fruit. Yield is reduced along with fruit quality. On cucumber, melon and watermelon, symptoms are milder. Control is by eliminating weeds. Resistance has been identified in several wild squash species (Provvidenti *et al.*, 1978a,b) but has not been transferred to cultivated species.

Watermelon mosaic virus

Watermelon mosaic virus (WMV), previously known as WMV-2, plagues squash, melon and other cucurbits in Europe and North America. It also attacks plants of other families. The virus causes stunting and mottling of the foliage, and unsightly green rings may develop on squashes normally having a yellow rind (Fig. 9.8).

WMV is transmitted mechanically and by aphids. Weed control reduces sources of inoculum. Application of insecticide or mineral oil curtails transmission by aphids. Aluminium mulch, rowcovers and other measures that discourage insect visitation are also used. 'Sweet Slice' cucumber, 'Freedom II' summer squash, the winter squash 'Redlands Trailblazer' (*C. maxima*), melon line B66-5, 'Cow Leg' bottle gourd and a watermelon relative, colocynth, are resistant to WMV.

Two of 16 wild/feral species of *Cucurbita* carried resistance to WMV (Provvidenti *et al.*, 1978a,b), one of which (*C. ecuadorensis*) was used to transfer resistance to *C. maxima* (Herrington *et al.*, 2001). A single dominant resistance gene (*Wmv*) has been identified in *C. moschata* (Brown *et al.*, 2003).

Zucchini yellow mosaic virus

Zucchini yellow mosaic virus (ZYMV) is a serious threat to cucurbits throughout the world today, although it was not recognized before 1980. The virus is

Fig. 9.8. The green rings on this yellow summer squash fruit were caused by watermelon mosaic virus. (Photo: J.R. Myers.)

Fig. 9.9. Zucchini yellow mosaic virus symptoms on winter squash. (A) 'Golden Delicious' fruit showing green patches that never ripen and reduce processed fruit quality. (B) Mosaic mottle symptoms on *C. maxima* leaves. (C) Close-up of leaf showing vein clearing symptoms.

very damaging to squash, melon and watermelon. Though generally less destructive to cucumber, ZYMV can still cause economic loss to this crop. It also attacks many other cucurbits.

ZYMV causes leaf mottling and extreme stunting (Fig. 9.9). Diseased fruit are often very distorted in shape and unmarketable. Squash cultivars normally having yellow fruit may have partially or entirely green fruit when infected. In processing types, green areas of the pericarp never ripen and will not blend into the purée. ZYMV is easily transmitted mechanically and by aphids.

Different strains and pathotypes of ZYMV exist. A single allele has been found in melon for resistance to one pathotype, but not to another. The *zym* allele discovered in a cucumber from Taiwan provides resistance to both of these pathotypes. 'Sweet Slice' and 'Taichung Mou Gua' cucumber cultivars are resistant. Resistance also occurs in bottle gourd.

In squash, resistance has been found in several species, including *C. equadorensis* and *C. moschata* (Provvidenti *et al.*, 1991), and these have been used to transfer resistance into *C. maxima* (Herrington *et al.*, 2001) and *C. pepo*, respectively (Paris and Cohen, 2000; Brown *et al.*, 2003). Resistant squash cultivars now include 'Jaguar', 'Tigress' and the transgenic 'Freedom II' (*C. pepo*); 'Nigerian Local', 'Menina', 'Soler' and 'Nicklow Delight' (*C. moschata*); and 'Redlands Trailblazer' (*C. maxima*). Multiple factors for resistance have been identified in *C. moschata* crosses (Pachner and Lelley, 2004; Pachner *et al.*, 2011). Pachner *et al.* (2011) identified five loci controlling resistance found in four cultivars. Three showed dominance and one recessive inheritance, while another dominant gene required complementation by a recessive pair for expression. The situation is similarly complex when introgressed into *C. pepo* (Formisano *et al.*, 2010), where the multigenic nature of resistance is compounded by interspecific effects on introgression. While a number of resistance genes have been discovered, these do not appear to be race specific.

10

INSECTS AND SPIDER MITES

Insects are often a major obstacle to the successful production of cucurbits (York, 1992). They can severely reduce yield, injure or kill plants, spread disease and adversely affect fruit quality. Spider mites can also be a problem, particularly with glasshouse cucumbers.

Integrated pest management requires an understanding of the biology and life history of the living organisms that interact with cucurbits. Management begins with cultural control methods, such as crop rotation, deep ploughing of crop residues, control of alternative hosts (i.e. weeds) and the use of trap crops. Because squash plants are strong attractors of the sweet potato whitefly, they can be planted as a trap crop to protect tomatoes (Coleman, 1995). Rowcovers are effective at restricting pest (e.g. cucumber beetle) access to young cucurbit plants. UV-reflective aluminium-coated plastic mulch discourages visitation by virus-carrying aphids. In the glasshouse, sticky boards and similar traps catch whiteflies and other flying insects.

Cultural and physical controls alone are rarely sufficient for cucurbit pest management. Foliar application of chemicals is the most common means of controlling insects and mites in the field and aerial fumigation is also used in the glasshouse. Soil fumigation, although expensive, can eradicate soilborne pests.

The trend in recent years has been to reduce the amount and spectrum of chemical pesticides applied to cucurbits. Instead of routinely applying insecticides on a periodic basis, as was often done previously, growers or consultants make inspections and count insects to determine if insecticide application is justified. In addition, 'biorational' pesticides have been developed that have less impact on the overall environment and non-target organisms. For example, insecticidal soaps and oils are less toxic chemicals for whitefly control than broad-spectrum preparations containing pyrethroids and organophosphates, although the latter generally are more effective at reducing pest populations. Biorational insecticides also include microbial-based preparations and pheromones that modify insect behaviour. The bacterium *Bacillus thuringiensis* attacks squash vine borers and other caterpillars, whereas the fungus

Verticillium lecanii is used in European glasshouses to infect whiteflies, thrips and aphids. Beneficial insect-attacking nematodes (e.g. *Steinernema carpocapsae*), applied to the soil or plant through watering or spraying, provide partial control for various cucurbit pests, including squash vine borers, cucumber beetles, cutworms and fruit fly larvae.

Beneficial insects are also employed for biological management. For example, the parasitic wasp *Encarsia formosa* can keep glasshouse whitefly populations in check. For cucumber, the recommended treatment is to release one parasite for every two plants while whitefly numbers are still low. A predator of spider mites, *Mesoseiulus longipes*, also works well in warm glasshouse environments. Aphid control for indoor and outdoor cucurbit crops can be achieved with *Aphidoletes aphidimyza* and various beneficial ladybird beetles, respectively. The concurrent use of chemical insecticides with biological controls needs to be carefully evaluated and monitored so that the beneficial insects and pollinators are not killed along with the pests.

Breeding cucurbits for insect and mite resistance has great potential, but has not progressed nearly as far as breeding for disease resistance. Few cucurbit cultivars have been bred for resistance to any insect, but sources of resistance are known (Robinson, 1992). Some landraces developed hundreds of years ago are resistant, though they were not intentionally bred for pest resistance.

An important problem in breeding cucurbits for insect resistance is that an allele for resistance to one insect may render the plant more susceptible to other insects. This is particularly true for genes influencing cucurbitacin content. These bitter and highly toxic compounds evidently evolved as a defence mechanism to protect cucurbits from insects and other herbivores. However, during the course of evolution, insects of the Chrysomelidae family developed mechanisms to tolerate cucurbitacins (Metcalf and Rhodes, 1990). Today, cucumber beetles and related insects, such as the red pumpkin beetle and corn rootworms, are not repelled by cucurbitacins like many other insects are, but instead are attracted to these compounds. Consequently, cucurbit plants that are bitterfree (lacking cucurbitacins) are resistant to cucumber beetles, but susceptible to spider mites, aphids, thrips and *Margaronia hylinata*.

CUCUMBER BEETLES

Cucumber beetles can cause severe damage and are the most important insect pest of cucurbits in many areas of the western hemisphere. *Acalymma vittatum* (Fig. 10.1) and several species of *Diabrotica* attack young seedlings, riddling the cotyledons and leaves with holes, damaging the roots, spreading diseases and killing plants by eating through the stem. Foliar damage reduces fruit set and lowers yield.

Cucumber beetles often appear in large numbers when cucurbit plants are still in the seedling stage, but are a problem at all stages of plant development.

Fig. 10.1. Adult striped cucumber beetle, *Acalymma vittatum.*

They may feed on blossoms, especially those of *Cucurbita maxima*, as well as on roots, fruit and foliage.

In addition to the extensive damage they cause by feeding, cucumber beetles are a major vector for bacterial wilt. The bacterium causing this disease overwinters in the intestines of cucumber beetles, and the beetles transfer the disease to cucurbits the following year. Cucumber beetles also transmit squash mosaic virus (SqMV) and other viruses.

The adult beetles are about 5 mm long and yellow, with black stripes or spots. They overwinter as unmated adults, under debris in fields or woodlands, emerging from hibernation in the spring when the temperature rises above 10°C. After mating, the female lays orange-yellow eggs around the base of cucurbits or other host plants. The larvae, which are white with black ends, hatch and burrow into the soil, where they feed on roots and underground stems. The white pupae persist in the soil for about a week. Then adults migrate above ground, where they feed on foliage. There may be two generations per year in warm climates and one in cool climates.

Several *Cucurbita pepo* cultivars, including 'Yellow Crookneck' and 'Acorn', have low cucurbitacin content in their cotyledons and are less preferred by cucumber beetles than cultivars such as 'Zucchini' and 'Caserta', which have more cucurbitacin. This difference in cucurbitacin content and attractiveness to cucumber beetles is conditioned primarily by a single gene (see 'Squash genes' in Chapter 3). Genetically controlled cucumber beetle resistance has also been reported in *C. moschata*, melon, watermelon and cucumber.

Cucurbitacins can be used as bait for cucumber beetles. A good source of cucurbitacins is the bitter 'Hawkesbury' watermelon. Bitter fruit of *C. maxima* ssp. *andreana* or another cucurbit are cut open and laced with an insecticide. Numerous cucumber beetles are lured to each cut fruit by its high cucurbitacin content and perish after eating the poison. The commercialization of poison baits containing cucurbitacin extracts is being investigated to make it easier to use.

Some insecticides control cucumber beetles. Autumn ploughing and other field sanitation measures may reduce pest population size in a field the following year, but sanitation and rotation alone may be inadequate control measures, since the adult beetles can emigrate from other fields.

CUCUMBER MOTH (CUCURBIT CATERPILLAR)

Cucumber moth (cucurbit caterpillar, *Diaphania indica*, formerly *Margaronia indica*) damages foliage and fruit of various cucurbits, including melon, in India, Thailand and other Asian countries. Fruit injured by this insect often rot prematurely. Resistance has been found in a primitive wild form of *Cucumis melo* previously known as *C. callosus*.

CUTWORMS

Cutworms (species of *Agrotis* and related genera) feed on and damage young cucurbits. The caterpillars hide in the soil during the day, coming out at night to chew on plant stems. Adults, measuring 10 mm long, and the morphologically similar but smaller nymphs overwinter in field debris. After mating, eggs are oviposited into leaves. There may be several generations per year. Cutworms can be controlled by baits and soil applications of insecticides. No resistance has been reported for cucurbits.

DARKLING GROUND BEETLES

Darkling ground beetles (*Blapstinus* spp.) may attack melons, especially if the fruit is cracked, making the fruit unmarketable. The beetles also feed on the netting of undamaged melons, and occasionally will feed on young stems. The small (10 mm long) blackish beetles live in the soil or hide in debris on the soil surface.

EPILACHNA BEETLES

Epilachna beetles (*Epilachna* spp.) are primarily pests of cucurbits and other crops in Africa. Watermelon, cucumber, summer squash and ivy gourd are affected. The adult beetles, measuring 6–8 mm long, are reddish to brownish yellow with black spots. Clusters of pale yellow eggs are laid on the underside of the leaf. The larvae and adults feed on the leaf, sometimes stripping the blade to the midrib. These insects also gnaw holes into stems and fruit. They are controlled with insecticidal sprays.

GLASSHOUSE WHITEFLY

Glasshouse whitefly (*Trialeurodes vaporariorum*), as its common name implies, is primarily a problem in the glasshouse, but it also inhabits cucurbit fields. Cucumber and other cucurbit plants can be stunted by glasshouse whiteflies sucking their plant juices. Heavily infested plants may become covered with a black sooty mould growing on the whitefly excreta. This pest can transmit melon yellows virus (MYV) to melon and cucumber.

Adult whiteflies reach 1.5 mm long and have pale yellow bodies and white wings. They prefer glasshouses and other warm moist areas with suitable plants from their wide range of hosts. Yellow-green eggs, measuring ca 0.5 mm long, are attached to the lower leaf surface by short stalks. The semi-transparent larvae, about 0.75 mm long, and the pupae also inhabit the lower leaf surface. Damage is done by sucking nymphs and adults.

Encarsia formosa is a predator of whiteflies and has been used to protect glasshouse tomato crops. However, the pubescence on cucumber leaves deters *E. formosa* from efficiently parasitizing whiteflies on cucumber. Consequently, the cucumber glabrous allele (*gl*) has been proposed as a means of biological control for the glasshouse whitefly. Undesirable pleiotropic effects of *gl* have discouraged the breeding of cultivars homozygous for this allele, but heterozygous plants appear normal except for trichome density being reduced sufficiently to permit *E. formosa* to reduce whitefly populations.

Resistance to the glasshouse whitefly is known in wild Cucumis species, but these species cannot be crossed with cucumber or melon. There are no resistant cultivars.

Insecticidal soap sprays are used to contain this pest. Because whiteflies are attracted to the colour yellow, squares of yellow plastic coated with a sticky substance are placed in glasshouses as traps and to monitor population size to determine if insecticide application is needed.

LEAFMINER

Leafminer (*Liriomyza* spp.) larvae burrow beneath the epidermis, especially on the upper surface of cucurbit leaves. They leave a trail of damaged tissue behind that externally appears to be circuitous lines of lighter green colour than the rest of the leaf. The chlorotic depressions in the leaves become targets for fungal infection. Leafminers may also spread viruses.

The adult female, a small fly about 1.5 mm long, oviposits into the leaf tissue, generally at the tip or side of the leaf. After feeding inside the leaf, the larvae emerge to pupate on the leaf surface, on another plant part or on the ground.

Leafminers are difficult to manage with insecticides once they are insulated between the leaf surfaces. Biorational control methods that do not destroy the natural enemies of leafminers are preferred. Cultivars with resistance to the vegetable leafminer, *Liriomyza sativae*, are available in melon, summer squash and pumpkin.

LEAF-FOOTED BUG

Leaf-footed bug (*Leptoglossus australis*) infests cucumber and other cucurbits in Australia, Africa, India, southeastern Asia and China. Adults and nymphs puncture stems, flowers and young fruit. Insect feeding results in dark spots on the fruit, which may drop prematurely. Heavy feeding causes stem tips to die.

The adults are 20–25 mm long and dark brown to grey with a pale orange stripe. Eggs laid in rows on the stem hatch in about a week, and five nymphal instars develop during the next 30 days.

Repeated insecticide applications may be needed to achieve control. Sanitation, including disking or removing vines after harvest, reduces the number of insects overwintering. A pheromone that attracts the male leaf-footed bug is used to monitor population density.

LEAFHOPPERS

Leafhoppers, including the melon leafhopper (*Empoasca abrupta*), the potato leafhopper (*E. fabae*) and several other species, attack cucurbits and many other plants. In western USA, the sugarbeet leafhopper (*Circulifer tenellus*) may transmit curly top virus when it feeds on cucurbits, although it prefers other hosts. Leafhoppers also spread the phytoplasma that causes aster yellows disease.

Adults jump or fly when disturbed, hence the name leafhopper. Leafhopper adults and nymphs feed on the lower leaf surface by sucking sap from the phloem. The adults, which are wedge-shaped, green and about 34 mm long, overwinter in warm climates and migrate to other areas in the spring. Eggs are oviposited into veins or petioles. Nymphs are similar in appearance to adults, but lack wings.

Various insecticides are used against leafhoppers; repeated applications are necessary for control, about every 1–2 weeks. At night, light traps can catch large numbers of individuals.

MELON APHID

Melon aphid (*Aphis gossypii*) attacks not only melon, but also cucumber, squash and other cucurbits. In large numbers, these insects reduce crop yield and quality by sucking the plant juices. Signs of heavy aphid infestation include curling of young leaves and blackened foliage or fruit, due to fungal growth stimulated by excreta of aphids. Frequently, even more serious losses are caused by melon aphids transmitting viral diseases. Those include cucumber mosaic virus (CMV), zucchini yellow mosaic virus (ZYMV), papaya ringspot virus (PRSV) and watermelon mosaic virus (WMV).

Aphids are small, sucking insects that often feed on the lower leaf surface or stem tips. They may be green, yellow or black; melon aphids are usually dark green. Both winged (alate) and wingless (apterous) forms occur for the same species. Winged forms can fly or be carried by the wind over long distances. Aphids overwinter as eggs in temperate regions. In the tropics, all stages of the life cycle may exist throughout the year. Aphids have a wide host range, which includes a number of different plant families. They can reproduce sexually or by parthenogenesis, quickly building up to large populations under favourable conditions. Often, numerous generations occur per year.

Some aphicides are systemics applied to the foliage or soil. Contact insecticides should target the undersides of leaves. Fungal preparations of *Verticillium lecanii* are effective in glasshouses when conditions are warm and humid.

Good weed control also helps to limit aphid populations. Early and late plantings of cucurbits should be geographically isolated, if possible, to minimize the spread of insects from one field to another. Brightly coloured aluminium mulch repels aphids and is most effective early in the season, before the mulch is covered by cucurbit vines. A genetic system to simulate the effect of aluminum mulch was proposed as a means of deterring aphids and preventing aphid-transmitted viruses by Oved Shifriss, who developed a summer squash line (NJ 260) with silvery leaves (due to gene *M* and modifiers).

Insect resistance is classified into three categories: non-preference (less attractiveness to insects), antibiosis (adverse effect on the insect feeding on that host) and tolerance (recovery from insect feeding). Each of these three factors has been determined to be involved in the aphid resistance of melon PI 371795 (Bohn *et al.*, 1972). This resistance is due to a single dominant allele, *Ag*. USDA researchers have backcrossed this allele into 'AR Hales Best Jumbo', 'AR Topmark' and 'AR 5'. Another allele (*Vat*), derived from Spanish melon landraces, reduces the amount of the virus transmitted by melon aphids. The *Vat*-resistant alleles have been used extensively to introgress into 100 Charentais cultivars being cultivated over 40% of melon cultivation areas in France. Genetic variation occurs in aphids for response to resistant genotypes of melon. 'Texas Resistant No. 1' melon was found to be resistant to one biotype of the melon aphid but susceptible to a second.

Aphid resistance has been reported for cucumber, summer squash, watermelon, luffa and wild *Cucumis* species. Cucumber allele bi increases susceptibility to aphids.

The green peach aphid, *Myzus persicae*, also attacks cucurbits, but prefers other host plants. Green peach aphids frequently migrate from one cucurbit plant to another while seeking a more preferred host, thereby spreading viruses. The pumpkin aphid, *Illinoia cucurbita*, occurs on squash and melon in western USA.

MELON FRUIT FLY

Melon fruit fly (*Dacus cucurbitae*) is one of the most serious pests of cucurbits in India. It also occurs in Japan, southeastern Asia, East Africa and Hawaii, attacking melon, cucumber, watermelon, squash and other cucurbits.

Adult flies are yellow-brown and 8–10 mm long. The female fly oviposits into immature fruit. The larvae hatch within a week and feed on the fruit, causing damage with their tunnels and providing entry for fruit-rotting organisms. The larvae grow to a length of 10–12 mm, then the yellowish-white maggots drop to the ground and pupate in the soil.

Sometimes, developing fruit can be covered to protect against fruit flies. For watermelon, dry grass mulching and paper wrapping are effective. Open paper bags are placed over bitter gourd fruit in some countries.

Sources of resistance have been found in *C. maxima*, bottle gourd, both domesticated species of luffa, bitter gourd, cucumber, melon and watermelon. Resistant melon fruit were found to have thicker, tougher rinds than melons susceptible to this pest. High silica content of the fruit rinds has also been associated with melon fruit fly resistance.

ORIENTAL FRUIT FLY

Oriental fruit fly (*Dacus dorsalis*) is closely related to the melon fruit fly. It is smaller, but otherwise looks like the melon fruit fly and has a similar life cycle. The oriental fruit fly has many more hosts than the melon fruit fly, including numerous fruit and vegetable crops in addition to cucurbits.

Oriental fruit fly is a serious pest in Hawaii and Asia. Stringent quarantine and eradication measures have been undertaken to prevent its introduction into California and Japan. Fumigation of fruit grown where the insect occurs is required before they can be shipped to quarantined areas. Baits that attract male flies are used to monitor the spread of the pest to new areas. When the flies are detected, eradication measures are quickly initiated.

Insects that parasitize the oriental fruit fly and sterilized male oriental fruit flies have been introduced for biological control. Other control measures include baits of sugar mixed with insecticide, removal of infested fruit and cultivation to disturb soilborne pupae.

PICKLEWORM

Pickleworm (*Diaphania nitidalis*) inhabits warm regions of southeastern USA, Mexico and the Caribbean islands. The larvae feed on blossoms and burrow into fruit of cucumber, melon and squash, reducing fruit set and yield and impairing fruit appearance. Secondary fruit rots invade damaged fruit. Similar damage is done by the melonworm, *D. hyalinata*, in southeastern USA. In India, *D. indica* is a pest on luffa.

Adult pickleworms and melonworms are about 20 mm long. The yellow-brown pickleworm moths and the silvery or white melonworm moths have a wing span of more than 2.5 cm. They fly at night and lay eggs on the lower surface of leaves, stems and fruit. Soon after hatching, pickleworm larvae tunnel into flowers and foliage and, later, burrow into fruit. Melonworm larvae generally feed on foliage. Young pickleworms are yellowish-white and young melonworms are greenish-yellow. They grow to 16–22 mm long and hibernate as pupae, a cocoon encased in silk and often rolled up within a leaf. There may be several generations per year, with different stages of development present at the same time.

Resistance has been found in cucumber, melon and squash. Glabrous (*gl*), a cucumber allele that prevents formation of trichomes on foliage, has been associated with resistance to pickleworms. Evidently, the adult female relies on the trichomes of the cucumber plant as a stimulus for oviposition, and lays fewer eggs on glabrous plants. The melonworm has a similar reaction to glabrous melon plants. *C. moschata* cultivars are generally more resistant than those of C. pepo.

In some areas, pickleworms and melonworms can be avoided by early planting. Insecticides need to be applied early for good control, before the larvae tunnel into the protective plant tissue. *Bacillus thuringiensis* provides only moderate control. Deep ploughing of crop residues to kill pickleworm larvae is recommended. *Melothria pendula* and bitter gourd, cucurbits that are weeds in Florida, are hosts for pickleworms and melonworms; their eradication can help to control insect populations on nearby crops.

RED PUMPKIN BEETLE

Red pumpkin beetle (*Aulacophora foveicollis*) is closely related to the cucumber beetle. It is one of the most significant cucurbit pests in India and also attacks plants in other Asian countries, the Middle East and southern Europe. This insect damages melon, luffa, bottle gourd, watermelon, squash and other cucurbits, particularly in the seedling to young plant stage. Adults feed on cotyledons, immature leaves and flowers.

The adult red pumpkin beetle is approximately 7 mm long and similar in appearance to the spotted cucumber beetle, except for its red-orange colour. The adult overwinters in the soil. In the spring, eggs are laid on the soil under debris. Later in the season oviposition may also be on the rinds of fruit touching the ground. Larvae pupate in the soil. Two or more generations occur per year.

The red pumpkin beetle is controlled with insecticides. Resistance has been found in melon, watermelon, bottle gourd, luffa and species of *Cucurbita*.

SPIDER MITES

Spider mites (*Tetranychus urticae* and several related species) are not insects, but arachnids, and they damage cucurbits by sucking plant juices. They cause distortion and chlorotic spotting of leaves, and can be recognized by their webs on the foliage. Spider mites can reduce yield and impair fruit quality. They are often a major problem for glasshouse cucumbers, and can also be serious pests in cucurbit fields, especially during hot, dry weather. They occur nearly everywhere that cucurbits are grown, and can be windblown to distant areas.

Spider mites, although quite small, can be seen. Their presence is also detected by their webs and the pale green, stippled appearance of infested leaves. Eggs are laid in large numbers on the lower leaf surface. Numerous generations may be produced in a single year. Overwintering in the field is usually as adults. Spider mites can migrate from grassy borders to cucurbit fields. Mowing and other disturbances that would stimulate this migration should be avoided. Acaricides are applied to cucurbits or fumigated in glasshouses to kill spider mites, but some spider mite populations may develop resistance to these chemicals. There are insect predators of spider mites, and killing these beneficial insects with insecticides such as pyrethroids may result in a build-up of spider mite populations. Resistance to spider mites has been reported in cucumber, melon and watermelon.

SQUASH BUG

Squash bug (*Anasa tristis*) prefers plants of *Cucurbita*, but in the absence of a squash host it may feed on cucumber, melon, watermelon or other cucurbits. It is distributed only in the western hemisphere, from Canada to South America.

Feeding by squash bugs causes leaves of squash plants to turn yellow, then black. Severe infestations cause the plants to wilt, a condition known as 'Anasa wilt'. Nymphs and adults may damage or kill squash vines by sucking plant juices from leaves and fruit. Squash bugs have been reported to transmit the bacterial wilt pathogen.

Adults are winged, dark brown and 17.5 mm long. Unmated adults overwinter, hibernating under dead vines and other protective debris. Large yellow eggs are laid on the lower surface of leaves of young squash plants early the next season. The eggs turn brown, then hatch into pale-coloured nymphs. The nymphs differ from the adults by being soft-bodied, wingless and initially green, later becoming grey and developing wing pads (Fig. 10.2). The nymphs moult four times. There is usually one generation per year in temperate regions.

Rotation and field sanitation after growing a squash crop help to control the squash bug. Adults and nymphs often hide under debris around the base of squash plants. Researchers have placed boards near the plants, as a means of monitoring the squash bug population and evaluating plants for resistance. Insecticides are more effective on young nymphs than on older nymphs and adults.

C. maxima is more susceptible to the squash bug than most cultivars of *C. pepo*, but butternut cultivars of *C. moschata* and cultigens of *C. argyrosperma* are highly tolerant.

SQUASH VINE BORER

Squash vine borer (*Melittia cucurbitae*) prefers squash, especially cultivars of *C. pepo*. Cucumber, watermelon, melon and other cucurbits can also be damaged, but the insect may not complete its life cycle on these hosts. Although the squash vine borer is restricted to the western hemisphere, from South America to Canada, the genus *Melittia* is distributed worldwide, and other species can attack squash species that are resistant to *M. cucurbitae*.

The orange and black moth, about 15 mm long, has opaque forewings and transparent rear wings with a wingspan of about 35 mm. The wasp-like moth lays small, reddish-brown eggs singly in the daytime at the base of squash

Fig. 10.2. Squash bug (*Anasa tristis*) nymphs on a squash leaf in a garden in Nebraska. (Photo: Kyle C. Broderick, University of Nebraska, Lincoln.)

plants, mostly on the stem or leaf petioles. The larvae hatch and tunnel into the stem of the plant. This often causes the stem to collapse, and the plant wilts and may die. The frass around the hole where the borer entered the stem is indicative of the insect's presence. The larvae, which are white with black head and thoracic shield when hatched, grow to 25 mm from an initial length of approximately 2 mm. After feeding within the stem for 2–4 weeks, the larvae fall to the ground and overwinter and pupate in the soil. Pupae are brown and about 16 mm long. There is one generation per year in temperate areas; additional generations may develop in regions with a longer growing season.

Home gardeners sometimes crush the eggs, mound soil about the stems of plants wounded by a borer in order to promote adventitious rooting, or kill borers with a sharp instrument. Commercial growers more often rely on insecticides, which are best applied to the base of plants early in the season. Control by insecticide may be difficult after eggs have hatched, since the borers are embedded in the stem, where they are protected from externally applied chemicals. Rotation is recommended to avoid planting in fields with soilborne pupae.

C. moschata and C. argyrosperma are resistant. It has been suggested that the squash vine borer resistance of C. moschata is associated with its hard, woody stem. Genetic differences in volatile chemicals of the host plant are also considered to influence oviposition.

SWEET POTATO WHITEFLY

Sweet potato whitefly (Bemisia tabaci) transmits viruses, including squash leaf curl virus (SLCV) and lettuce infectious yellow virus (LIYV), to squash and melon. It also damages summer squash by causing silver leaf disorder, which has become a serious problem in recent years in Florida, France and the Middle East. Affected plants have prominent leaf veins and a silvery appearance on the upper leaf surface. Fruit colour may become pale and yield is often reduced. Silvering has been attributed to a toxic reaction to feeding by the nymphs. Summer squash cultivars differ in sensitivity to the toxic factor. In the 1990s, an especially troublesome population (the poinsettia strain) of this insect built up to high populations in the Imperial Valley of California, causing severe damage to melon and other crops despite attempts to control it with pesticides. Cucurbita lundelliana is reportedly resistant to the sweet potato whitefly.

THRIPS

Thrips, including the western flower thrip (Frankliniella occidentalis), the sugar beet thrip (Heliothrips femoralis) and other species, are destructive to melon, cucumber (especially in glasshouses) and other cucurbits. Although thrips are so small (1 mm long) as to be barely visible, they can occur in such large numbers that considerable damage is done. They frequently inhabit flowers, but also feed on foliage. Affected leaves have silvery areas with black, necrotic spots and margins that curl downward. Thrips are important vectors of tospoviruses. Weed management and insecticides are common control measures. Resistance to western flower thrips has been reported for cucumber. Minute pirate bugs can be important in controlling thrips populations.

WIREWORMS

Wireworms live in the soil and feed on the roots of cucurbits and other plants. Damage is likely to be most severe with early plantings on light soils. Wireworms are the immature stages of click beetles, and the adult stage is harmless to cucurbits. Larvae are yellow to brown, measuring 25 mm long or less.

Pests in Midwestern USA are the corn wireworm (*Melanotus cribulosus*), sugarbeet wireworm (*Limonius californicus*) and wheat wireworm (*Agriotes mancus*). Pests in Western USA are Pacific coast wireworm (*Limonius canus*), sugarbeet wireworm (*L. californicus*), western field wireworm (*L. infuscatus*), Columbia basin wireworm (*L. subauratus*) and great basin wireworm (*Ctenicera pruinina*).

11

ABIOTIC STRESSES

Numerous abiotic stresses impact yield, quality and adaptation of cucurbit crops under the diverse cultivation systems across the globe. Irrigation systems, elevation, soil type, solar radiation, wind, pathogens, fertility and genotype all interact to determine the degree to which abiotic stress may impact the crop. Dramatic improvements in irrigation technology along with protected culture practices have helped to reduce drought, wind, heat and cold stress, but often at a high price. In addition, availability of crop land in regions with the best soils, topography, elevation and temperatures is limited by competition from more valuable crops or other human activities. As a result, cucurbits are often produced in regions with drought, salinity, temperature and nutrient stress. In subsistence agriculture systems, where cucurbits are often a valuable source of carbohydrates, protein and micronutrients (β-carotene, vitamin E), expensive inputs such as precision irrigation, shading, tunnels and grafting are unrealistic solutions for abiotic stress reduction. The best approach in these areas and across all cropping systems is genetic stress resistance. This has been achieved in many agronomic crops and progress is being made in many vegetable crops.

Tolerance or resistance to abiotic stress factors has been documented in the germplasm of multiple cucurbit crops. Utilizing wild accessions, landraces and known resistance sources, genes for abiotic stress resistance are being identified and introgressed into cultivars with desirable fruit quality. Complete genome sequences and genetic maps for many major cucurbits now exist, allowing detection of important stress resistance genes. Dramatic improvements in high-throughput molecular marker screening will expedite the deployment of these genes in commercially acceptable cultivars. Another approach has been to utilize sequence information for known stress resistance genes from *Arabidopsis* to explore the cucurbit genomes for homologous sequences. This will allow exploitation of endogenous resistance genes rather than utilizing genetic transformation to introduce transgenes. With the new gene editing technology, it may be possible to modify these endogenous stress resistance genes to enhance their effects in the near future. Just as deployment of major disease resistance genes has allowed growers to reduce pesticide applications, exploiting major stress resistance genes will contribute to reduced water and fertilizer inputs. This should enhance sustainability of cucurbit cropping systems at a time when resources are becoming limited or more expensive for growers.

BITTER FRUIT

Bitter fruit are produced in most cucurbit species as the result of genetics and environment. The bitter compounds are cucurbitacins that can occur in the foliage and fruit. Genes have been identified that control the bitter compounds. For example, the *bi-1* and *bi-2* genes in cucumber each produce bitterfree foliage in homozygous recessive condition. Also, dominant alleles of the *Bt-1* and *Bt-2* genes each produce bitter fruit.

The environment will also cause fruit to be bitter, especially if the fruit develop during a dry period. This is common with cucumber and can be remedied by irrigating during fruit set, or growing cultivars that have bitterfree foliage. Insects pests do much more leaf damage on bitterfree cucumbers and so that type is usually grown in protected culture, such as in greenhouses. Fruit bitterness may also be caused by high temperature, or wide changes in temperature.

Squash seeds should be produced with adequate isolation distance (2 km) from gourds of the genus *Cucurbita*, since squash will cross with some gourds. Bitter fruit can result from the accidental cross of gourd during seed production of a squash cultivar.

Fig. 11.1. Chilling tolerant vs. chilling sensitive cucumber accessions at the seedling stage.

COLD (CHILLING TOLERANCE)

Cold-tolerant cucumbers and melons have been developed using plant breeding methods. In muskmelon, Hutton and Loy (1992) identified a cold-temperature germination trait. It was determined that both recessive genes and a cytoplasmic factor were involved. Subsequently, they developed lines with cold-germination ability for production regions with sub-optimal soil temperatures during field seeding. Mature plant cold tolerance in melon demonstrated the positive impact of heterosis on performance traits under lower than normal temperatures. F_1 hybrids were produced by intercrossing seven open-pollinated cultivars with above-average cold tolerance. Many of the hybrids showed transgressive segregation, outperforming the parents under two regimes of cold temperatures.

Temperatures below 10°C cause chilling injury in cucurbits. Watermelon and squash have more chilling tolerance than luffa gourd and melon, which have more chilling tolerance than cucumber. Chilling injury is greater under the following conditions: (i) low temperatures (below 10°C); (ii) longer chilling duration (greater than 7 h); (iii) higher light intensity during chilling; (iv) high air speed during chilling; and (v) higher growth temperature before chilling. Accessions have been identified with chilling tolerance and there are genes for tolerance in the cytoplasm (causing maternal inheritance) as well as the nucleus (single gene inheritance) (Fig. 11.1). Thus, a chilling-tolerant hybrid could be developed using a female parent with a cytoplasmic gene and a male parent with a dominant nuclear gene. That situation exists in cucumber, since it has moderate resistance from a cytoplasmic gene and the *Ch* dominant nuclear gene for chilling tolerance.

DROUGHT (WATER DEFICIT)

Drought during the production of cucurbit crops can lead to shorter vines, cause delayed flowering, shift the plant towards maleness (with more staminate and fewer pistillate flowers) and reduce fruit yield and quality.

Melon and watermelon are important crops in warm climates, often in regions that have problems with drought and salinity. Screening studies of watermelon germplasm by researchers in Turkey, Japan and the USA have revealed genetic variation for drought tolerance. Accessions of wild species of *Citrullus* were identified that had drought tolerance and that produced elevated levels of citrulline. Researchers identified a gene for high citrulline (Kawasaki *et al.*, 2000). There were 25 *Citrullus lanatus* accessions (originating from Africa) of 1066 germplasm and breeding lines tested that had high drought tolerance in the seedling stage (Zhang *et al.*, 2011). Many of those accessions were from Zimbabwe and about half were sweet watermelon (*Citrullus lanatus*), with the others being citron (*Citrullus amarus*). These have potential for breeding both scion and rootstock cultivars having drought tolerance.

Research at Texas A&M University has investigated root vigour and morphology in crosses of sweet watermelon ('Crimson Sweet' and 'Dixielee') with citron (*Citrullus amarus*), as well as western shipper (Cantalupensis) with Inodorus melon. Heterosis for root length, area, diameter and lateral numbers was observed. In field trials some of the derived breeding lines had drought tolerance and resistance to vine decline (Monosporascus root rot). Researchers in Turkey screened 85 watermelon accessions for drought tolerance in field experiments with deficit irrigation. More than one-third demonstrated drought tolerance based on nine traits, with one accession rating 99 on a scale of 0–100 (Karipcin *et al.*, 2008). The drought-tolerant accessions could be used to develop new cultivars. Small-fruited watermelon may have improved drought tolerance relative to large-fruited ones.

FLOODING (EXCESS WATER)

Cucurbit crops are sensitive to flooding, which is why they are mostly produced on well-drained soils, or in arid regions. Raised beds are useful in areas with rain during the production season, unless the soils are sandy. Flooding tolerance is important in humid regions such as southeast Asia and Brazil, but prolonged periods of heavy rain cause reduced fruit yield and quality.

Excess water causes reduced oxygen in the soil (hypoxia). Some cucurbits, such as bottle gourd, have tolerance to hypoxia in the roots during flooding. Bottle gourd is useful as a rootstock for bitter gourd, watermelon and other flooding-sensitive cucurbits. Research on flooding tolerance among cucumber accessions has identified the importance of adventitious roots to resist hypoxia during flooding (Xu *et al.*, 2016). The most flood-tolerant accessions had numerous adventitious roots in the hypocotyl region, while susceptible accessions did not. Major genes controlling adventitious root development have been identified in several crops, so the development of molecular markers for such genes in cucurbits should be possible.

Excess water can also reduce the nutrient content of the soil, reducing plant growth as a result. Cucurbits growing in flooded soil may be chlorotic and stunted. In response, aerial (adventitious) roots often develop at the base of the plant in cucumber and melon, but not in watermelon and squash.

HEAT

Heat tolerance is an important trait for sustainable vegetable production, considering that many countries in the tropics have high temperatures in the crop production season. Although many cucurbits originated in tropical regions, not all are capable of producing a reliable crop in those areas, especially in the warmer growing periods. Humans have improved the adaptation of cultivars to high temperature for many of the cultivated cucurbits. However, continued efforts are needed to improve fruit yield and quality. Fruit quality of even the more heat-tolerant cultivars is often inferior to that of the same crops grown during seasons where temperatures are closer to optimum. Flower abortion and pollen sterility are common symptoms of heat stress in melon and cucumber. Reduced fruit size is another common response to high temperature. Additionally, sunburn of fruit and desiccation of leaves may occur.

Cucurbits grow well in tropical climates, since they are warm-season vegetable crops. However, they grow poorly at high temperatures (38–45°C), where growth is slow and leaf margins become chlorotic. At 42–45°C, young leaves become yellow-green, flowers abort, sex expression changes to more staminate and yield is reduced.

Dudaim and Agrestis groups of *Cucumis melo* have demonstrated heat tolerance in test locations around Texas, while Cantalupensis and Momordica types were not able to tolerate high temperatures (e.g. 40°C). Inodorus types varied in response, with some honeydew cultivars exhibiting reduced fruit set and reduced sunburn, compared with others. The cultivars 'TAM Perlita' and 'TAM Dew Improved' were developed at Weslaco, Texas, and have demonstrated improved heat tolerance, as well as fruit setting abilities under high heat and humidity relative to other cultivars.

NUTRIENT DEFICIENCIES AND EXCESSES

Calcium deficiency causes blossom end rot. It occurs most often in watermelon, squash and pumpkin. Low calcium availability can be due to low soil calcium, drought before fruit set, or high nitrogen fertility that causes rapid vine growth and soil calcium depletion. Some watermelon cultivars are more susceptible to blossom end rot, especially those with elongate fruit.

Manganese toxicity is associated with acid soils. It is more serious in watermelon and melon than in cucumber, squash and pumpkin. Symptoms are observed before harvest, with the leaves near the crown becoming pale green. The problem can be managed by raising soil pH using lime (applied in the previous year).

Nitrogen deficiency symptoms include yellowing of older leaves on the plant, slow growth, thin stems and distorted fruit. Nitrogen excess results in rapid growth of vines, especially branches, tendrils and leaves, and can make the plant more susceptible to stressful conditions, as well as insect feeding.

Phosphorus deficiency symptoms include stunted plants, dark green leaves and brown patches between the leaf veins. The patches spread and the leaf dies prematurely. Fruit set is reduced.

Potassium deficiency is especially a problem in cucumber compared with other cucurbits. Symptoms include stunted plants that have short internodes and small leaves. Older leaves become yellow, starting at the leaf margin, and fruit develop abnormally.

POLLINATION PROBLEMS

Cucurbits grown in the field require pollination for proper fruit set. Many cucurbits have staminate (male) and pistillate (female) flowers on the same plant and are self-fertile. However, in the case of seedless (triploid) watermelon, all plants of the hybrid are male and female sterile. In order to set fruit on the triploid, a diploid pollenizer must be planted in the field, and pollinating insects should be provided. Usually, one row of diploid pollenizer (producing seeded fruit) will be provided per three to five rows of triploid seedless plants. Alternatively, the diploid can be planted in the same row as the triploid, often one diploid to three or four triploid plants. In this case, the diploid is often a small-fruited type (< 1 kg per fruit) that is not harvested.

In the case of gynoecious (all female) cucumber, a monoecious pollenizer is included in the seed packet (as a blend) to provide pollen, often at the rate of 12–15% of the plants in the field. Thus, gynoecious cultivars are actually not a single cultivar, but a cultivar blend of a gynoecious and a monoecious hybrid.

Inadequate pollination may result in fruit defects. Too little pollen on the stigmas of the pistillate flower may result in nubbin (short and pointed) or bottleneck (constricted fruit diameter at the stem end, especially in watermelon) fruit. Too little pollen on one of the three stigmas of the pistillate flower may result in crooked (curved) fruit. Pollination problems can be reduced by putting two to three hives of honey bees in each hectare of cucurbit crop being produced, along with a pollenizer (monoecious cucumber or diploid watermelon). Too few diploid (pollenizer) watermelon plants in a field of triploid (seedless) watermelon may result in fruit with internal defects, including bottleneck, triangular cross-section fruit and hollow heart. Pollination problems occur more in watermelon cultivars having fruit that are elongate (rather than round or oval) in shape.

Cucurbits are less preferred by honey bees than other crops and so it may help to grow cucurbits together, rather than with other crops, such as clover. Insecticides should be applied in the afternoon when honey bees are less active in the field. This will avoid harm to the pollinating insects.

SALINITY

Melon cultivars screened under saline conditions showed genetic differences in tolerance. Salinity tolerance is heritable, since progress was made in breeding programmes in Israel and Egypt (Shannon and Grieve, 1999). However, salinity tolerance was not found to be due to exclusion of ions or accumulation in specific tissues. Also, different stages of growth varied in expression of salinity tolerance. In cucumber, the gene *sa* from PI 192940 was found to provide tolerance to salinity.

Appendix: Common Cucurbit Names and their Scientific Name Equivalents

SORTED BY COMMON NAME

Common name	Scientific equivalent(s)
Acorn squash	Particular cultivars of *Cucurbita pepo*
African horned melon	*Cucumis metuliferus*
Angled luffa	*Luffa acutangula*
Antidote vine	*Fevillea cordifolia*
Balsam apple	*Momordica balsamina*
Banana squash	Particular cultivars of *Cucurbita maxima*
Bell squash	Particular cultivars of *Cucurbita moschata*
Bitter gourd	*Momordica charantia*
Bottle gourd	*Lagenaria siceraria*
Bryony	Various species of *Bryonia*
Buffalo gourd	*Cucurbita foetidissima*
West Indian or burr gherkin	*Cucumis anguria*
Cantaloupe	Particular cultivars of *Cucumis melo*
Casabanana	*Sicana odorifera*
Chayote	*Sechium edule*
Cheese pumpkin	Particular cultivars of *Cucurbita moschata*
Chinese snake gourd	*Trichosanthes kirilowii*
Citron	*Citrullus lanatus* var. *citroides*
Cochinchin gourd	*Momordica cochinchinensis*
Cocozelle	Particular cultivars of *Cucurbita pepo*
Colocynth	*Citrullus colocynthis*
Crookneck	Particular cultivars of *Cucurbita pepo, C. moschata*
Cucumber	*Cucumis sativus*
Cushaw	Particular cultivars of *Cucurbita argyrosperma, C. moschata*
Delicious squash	Particular cultivars of *Cucurbita maxima*
Egusi	*Citrullus lanatus* var. *citroides; C. colocynthis; Cucumeropsis mannii*

Continued

Common name	Scientific equivalent(s)
Fig leaf gourd	*Cucurbita ficifolia*
Fluted pumpkin	*Telfairia occidentalis*
Gherkin	*Cucumis anguria;* particular cultivars of *C. sativus*
He-zi-cao	*Actinostemma tenerum*
Hubbard	Particular cultivars of *Cucurbita maxima*
Indreni	*Trichosanthes lepiniana*
Ivy gourd	*Coccinia grandis*
Japanese snake gourd	*Trichosanthes ovigera*
Jiao-gu-lan	*Gynostemma pentaphyllum*
Kaksa	*Momordica dioica*
Lard plant	*Hodgsonia macrocarpa*
Lollipop climber	*Diplocyclos palmatus*
Luffa, or loofah	Species of *Luffa*
Luo-guo-di	*Hemsleya amabilis*
Luo-han-guo	*Siraitia grosvenorii*
Malabar gourd	*Cucurbita ficifolia*
Marrow	Particular cultivars of *Cucurbita maxima, C. pepo*
Melon	*Cucumis melo*
Mi-mao-gua-lou	*Trichosanthes villosa*
Muskmelon	Particular cultivars of *Cucumis melo*
!nara	*Acanthosicyos horridus*
Ornamental gourd	Particular cultivars of *Cucurbita pepo*
Oyster nut	*Telfairia pedata*
Pickling melon	Particular cultivars of *Cucumis melo*
Pointed gourd	*Trichosanthes dioica*
Preserving melon	*Citrullus lanatus* var. *citroides*
Pseudo-fritillary	*Bolbostemma paniculatum*
Pumpkin	Particular cultivars of *Cucurbita pepo, C. maxima, C. moschata, C. argyrosperma*
Red hail stone	*Thladiantha dubia*
Round melon	*Praecitrullus fistulosus*
Scallop squash	Particular cultivars of *Cucurbita pepo*
Show pumpkin	Particular cultivars of *Cucurbita maxima*
Smooth luffa	*Luffa cylindrica*
Snake gourd	*Trichosanthes cucumerina*
Snake melon	Particular cultivars of *Cucumis melo*
Sponge plant	*Momordica angustisepala*
Squash	Species of *Cucurbita*, particularly *C. argyrosperma, C. maxima, C. moschata, C. pepo*
Squirting cucumber	*Ecballium elaterium*
Straightneck	Particular cultivars of *Cucurbita pepo*
Stuffing cucumber	*Cyclanthera pedata*
Summer squash	Particular cultivars of *Cucurbita pepo*

Continued

Common name	Scientific equivalent(s)
Teasel gourd	*Cucumis dipsaceus*
Tinda	*Praecitrullus fistulosus*
Turban squash	Particular cultivars of *Cucurbita maxima*
Vegetable marrow	Particular cultivars of *Cucurbita pepo*
Watermelon	*Citrullus lanatus*
Wax gourd	*Benincasa hispida*
White-seeded melon	*Cucumeropsis mannii*
Wild cucumber	*Echinocystis lobata*
Winter melon	Particular cultivars of *Cucumis melo*; *Benincasa hispida*
Winter squash	Particular cultivars of *Cucurbita argyrosperma, C. maxima, C. moschata, C. pepo*
Xishuangbanna gourd	*Cucumis sativus* var. *xishuangbannesis*
Zucchini	Particular cultivars of *Cucurbita pepo*

SORTED BY SCIENTIFIC NAME

Scientific name	Common name
Acanthosicyos horridus	!nara
Actinostemma tenerum	He-zi-cao
Benincasa hispida	Wax gourd, winter melon
Bolbostemma paniculatum	Pseudo-fritillary
Bryonia spp.	Bryony
Citrullus colocynthis	Colocynth, egusi
Citrullus lanatus	Watermelon
Citrullus lanatus var. *citroides*	Citron, egusi, preserving melon
Coccinia grandis	Ivy gourd
Cucumeropsis mannii	White-seeded melon, egusi
Cucumis anguria	West Indian or burr gherkin
Cucumis dipsaceus	Teasel gourd
Cucumis melo	Melon
Cucumis metuliferus	African horned melon
Cucumis sativus	Cucumber
Cucumis sativus var. *xishuangbannesis*	Xishuangbanna gourd
Cucurbita argyrosperma	Squash, pumpkin
Cucurbita ficifolia	Malabar gourd, fig leaf gourd
Cucurbita foetidissima	Buffalo gourd
Cucurbita maxima	Squash, pumpkin
Cucurbita moschata	Squash, pumpkin
Cucurbita pepo	Squash, pumpkin, gourd
Cyclanthera pedata	Stuffing cucumber

Continued

Scientific name	Common name
Diplocyclos palmatus	Lollipop climber
Ecballium elaterium	Squirting cucumber
Echinocystis lobata	Wild cucumber
Fevillea cordifolia	Antidote vine
Gynostemma pentaphyllum	Jiao-gu-lan
Hemsleya amabilis	Luo-guo-di
Hodgsonia macrocarpa	Lard plant
Lagenaria siceraria	Bottle gourd
Luffa acutangula	Angled luffa
Luffa cylindrica	Smooth luffa
Luffa spp.	Luffa
Momordica angustisepala	Sponge plant
Momordica balsamina	Balsam apple
Momordica charantia	Bitter gourd
Momordica cochinchinensis	Cochinchin gourd
Momordica dioica	Kaksa
Praecitrullus fistulosus	Round melon, tinda
Sechium edule	Chayote
Sicana odorifera	Casabanana
Siraitia grosvenorii	Luo-han-guo
Telfairia occidentalis	Fluted pumpkin
Telfairia pedata	Oyster nut
Thladiantha dubia	Red hail stone
Trichosanthes cucumerina	Snake gourd
Trichosanthes dioica	Pointed gourd
Trichosanthes kirilowii	Chinese snake gourd
Trichosanthes lepiniana	Indreni
Trichosanthes ovigera	Japanese snake gourd
Trichosanthes villosa	Mi-mao-gua-lou

REFERENCES

Achigan-Dako, E.G. (2008) *Phylogenetic and genetic variation analyses in cucurbit species (Cucurbitaceae) from West Africa: definition of conservation strategies.* Cuvillier Verlag, Gottingen.

Achigan-Dako, E.G., Fuchs, J., Ahanchede, A. and Blattner, F.R. (2008) Flow cytometric analysis in *Lagenaria sacraria* (Cucurbitaceae) indicates correlation of genome size with usage types and growing elevation. *Plant Systematics and Evolution* 276, 9–19.

Adams, P., Graves, C.J. and Winsor, G.W. (1992) Some responses of cucumber, grown in beds of peat, to N, K and Mg. *Journal of Horticultural Science* 67, 877–884.

Akoroda, M.O., Ogbechie-Odiaka, N.I., Adebayo, M.L., Ugwo, O.E. and Fuwa, B. (1990) Flowering, pollination and fruiting in fluted pumpkin (*Telfairia occidentalis*). *Scientia Horticulturae* 43, 197–206.

Anderson, A.P. (1894) The grand period of growth in fruit of *Cucurbita pepo* determined by weight. *Minnesota Botanical Series* 1, 238–279.

Andeweg, J.M. and Bruyn, J.W. de (1958) Breeding of non-bitter cucumbers. *Euphytica* 7, 13–20.

Ando, K., Carr, K.M. and Grumet R. (2012) Transcriptome analyses of early cucumber fruit growth identifies distinct gene modules associated with phases of development. *BMC Genomics* 13, 518.

Andres, T.C. (1990) Biosystematics, theories on the origin, and breeding potential of *Cucurbita ficifolia*. In: Bates, D.M., Robinson, R.W. and Jeffrey, C. (eds) *Biology and Utilization of the Cucurbitaceae.* Cornell University Press, Ithaca, New York, pp. 102–119.

Anonymous (2018) Vegetable grafting research-based information portal. Available at: http://www.vegetablegrafting.org/resources/reference-database/ (accessed 9 July 2018).

Arce-Ochoa, J.P., Dainello, F., Pike, L.M. and Drews, D. (1995) Field performance comparison of two transgenic summer squash hybrids to their parental hybrid line. *HortScience* 30, 492–493.

Aung, L.H., Ball, A. and Kushad, M. (1990) Developmental and nutritional aspects of chayote (*Sechium edule*, Cucurbitaceae). *Economic Botany* 44, 157–164.

Azevedo-Meleiro, C.H. and Rodriguez-Amaya, D.B. (2007) Qualitative and quantitative differences in carotenoid composition among *Cucurbita moschata*, *Cucurbita maxima*, and *Cucurbita pepo*. *Journal of Agricultural and Food Chemistry* 55, 4027–4033.

Baird, J.R. and Thieret, J.W. (1988) The bur gherkin (*Cucumis anguria* var. *anguria*, Cucurbitaceae). *Economic Botany* 42, 447–451.

Bang, H., Davis, A.R., Kim, S., Leskovar, D.I. and King, S.R. (2010) Flesh color inheritance and gene interactions among canary yellow, pale yellow, and red watermelon. *Journal of the American Society for Horticultural Science* 135, 362–368.

Behera, T.K., Rao, A.R., Amarnath, R. and Kumar, R.R. (2016) Comparative transcriptome analysis of female and hermaphrodite flower buds in bitter gourd (*Momordica charantia L.*) by RNA sequencings. *Journal of Horticultural Science and Biotechnology* 91, 250–257.

Benzioni, A., Mendlinger, S., Ventura, M. and Huyskens, S. (1993) Germination, fruit development, yield, and postharvest characteristics of *Cucumis metuliferus*. In: Janick, J. and Simon, J.E. (eds) *New Crops*. John Wiley and Sons, New York, pp. 553–557.

Bharathi, L.K., Munshi, A.D., Vinod, Chandrashekaran, S., Behera, T.K., Das, A.B., John, K.J. and Vishalnath (2011) Cytotaxonomical analysis of *Momordica* L. (Cucurbitaceae) species of Indian occurrence. *Journal of Genetics* 90, 21–30.

Bharathi, L.K., Joh, J.K., Singh, H.S., Srinivas, P. and Sivakumar, P.S. (2015) Tapping edible and medicinal potential of sweet gourd. *Indian Horticulture* May–June, 9–11.

Bhave, M.R., Gupta, V.S. and Ranjekar, P.K. (1986) Arrangement and size distribution of repeat and single copy DNA sequences in four species of Cucurbitaceae. *Plant Systematics and Evolution* 152, 133–151.

Blancard, D., Lecoq, H. and Pitrat, M. (1994) *Colour Atlas of Cucurbit Diseases: Observation, Identification and Control*. John Wiley and Sons, New York.

Bo, K.L., Song, H., Shen, J., Qian, C.T., Staub, J.E., Simon, P.W., Lou, Q.F. and Chen, J.F. (2012) Inheritance and mapping of the *ore* gene controlling the quantity of β-carotene in cucumber (*Cucumis sativus* L.) endocarp. *Molecular Breeding* 30, 335–344.

Bohn, G.W., Kishaba, A.N. and Toba, H.H. (1972) Mechanisms of resistance to melon aphid in a muskmelon line. *HortScience* 7, 281–282.

Bonina-Noseworthy, J., Loy, J.B., Curran-Celentano, J., Sideman, R. and Kopsell, D.A. (2016) Carotenoid concentration and composition in winter squash: variability associated with different cultigens, harvest maturities, and storage times. *HortScience* 51, 472–480.

Brown, R.N. and Myers, J.R. (2002) A genetic map of squash (*Cucurbita* sp.) with randomly amplified polymorphic DNA markers and morphological markers. *Journal of the American Society for Horticultural Science* 127(4), 568–575.

Brown, R.N., Bolanos-Herrera, A., Myers, J.R. and Jahn, M.M. (2003) Inheritance of resistance to four cucurbit viruses in *Cucurbita moschata*. *Euphytica* 129, 253–258.

Bruton, B.D., Fish, W.W., Roberts, W. and Popham, W. (2009) The influence of rootstock selection on fruit quality attributes of watermelon. *Open Food Science Journal* 3, 15–34.

Cai, Y., Luo, Q., Sun, M. and Corke, H. (2004) Antioxidant activity and phenolic compounds of 112 traditional Chinese medicinal plants associated with anticancer. *Life Sciences* 74, 2157–2184.

Cantliffe, D.J. and Phatak, S.C. (1975) Plant population studies with pickling cucumbers grown for once-over harvest. *Journal of the American Society for Horticultural Science* 100, 464–466.

Castetter, E.F. (1930) Species crosses in the genus *Cucurbita*. *American Journal of Botany* 17, 41–57.

Chandrasekaran, J., Brumin, M., Wolf, D., Leibman, D., Klap, C., Pearlsman, M., Sherman, A., Arazi, T. and Gal-on, A. (2016) Development of broad virus resistance in non-transgenic cucumber using CRISPR/Cas9 technology. *Molecular Plant Pathology* 17, 1140–1153.

Chee, P.P. and Slightom, J.L. (1991) Transfer and expression of cucumber mosaic virus coat protein gene in the genome of *Cucumis sativus*. *Journal of the American Society of Horticultural Science* 116, 1098–1102.

Chen, J.F., Staub, J.E., Tashiro, Y., Isshiki, S. and Miyazaki, S. (1997) Successful interspecific hybridization between *Cucumis sativus* L. and *C. hystrix* Chakr. *Euphytica* 96, 413–419.

Chen, J., Staub, J., Qian, C., Jiang, J., Luo, X. and Zhuang, F. (2003) Reproduction and cytogenetic characterization of interspecific hybrids derived from *Cucumis hystrix* Chakr. × *Cucumis sativux* L. *Theoretical and Applied Genetics* 106, 688–695.

Chen, Q., Ma, C., Qian, J., Lan, X., Chao, N., Sun, J. and Wu, Y. (2016) Transcriptome sequencing of *Gynostemma pentaphyllum* to identify genes and enzymes involved in triterpenoid biosynthesis. *International Journal of Genomics* 2016, 7840914. DOI: 10.1155/2016/7840914

Chomicki, G. and Renner, S.S. (2014) Watermelon origin solved with molecular phylogenetics including Linnaean material: another example of museomics. *New Phytologist* 205, 526–532.

Christopher, D.A. (1982) Influence of foliarly applied growth regulators on sex expression in watermelon. *Journal of the American Society for Horticultural Science* 107, 401–404.

Cohen, R., Hanan, A. and Paris, H.S. (2003) Single-gene resistance to powdery mildew in zucchini squash (*Cucurbita pepo*). *Euphytica* 130, 433–441.

Cohen, R., Burger, Y., Horev, C., Koren, A. and Edelstein, M. (2007) Introducing grafted cucurbits to modern agriculture: the Israeli experience. *Plant Disease* 91, 916–923.

Coleman, R.G. (1995) Methods that help manage incidence of whitefly-transmitted virus. *Florida Grower and Rancher* 88(12), 41.

Colle, M., Straley, E.N., Makela, S.B., Hammar, S.A. and Grumet, R. (2014) Screening the cucumber plant introduction collection for young fruit resistance to *Phytophthora capsici*. *HortScience* 49, 244–249.

Cui, H. and Zhang, X. (1991) Cucumber cultivar improvement in the People's Republic of China. *Cucurbit Genetics Cooperative Report* 14, 5–7.

Culpepper, C.W. and Moon, H.H. (1945) Differences in the composition of the fruits of *Cucurbita* varieties at different ages in relation to culinary use. *Journal of Agricultural Research* 71, 111–136.

Cummings, M.B. and Jenkins, E.W. (1925) Hubbard squash in storage. *Bulletin of the Vermont Agricultural Experiment Station* 251, 1–35.

Darwin, C. (1906) *The Movements and Habits of Climbing Plants*. John Murray, London.

Davies, J.N. and Kempton, R.J. (1976) Some changes in the composition of the fruit of the glasshouse cucumber (*Cucumis sativus*) during growth, maturation and senescence. *Journal of the Science of Food and Agriculture* 27, 413–418.

Davis, J.M. (1994) Luffa sponge gourd production practices for temperate climates. *HortScience* 29, 263–266.

Deakin, J.R., Bohn, G.W. and Whitaker, T.W. (1971) Interspecific hyridization in *Cucumis*. *Economic Botany* 25, 195–211.

Decker, D.S. (1986) A biosystematic study of *Cucurbita pepo*. PhD thesis, Texas A&M University, College Station, Texas.

Decker, D.S. (1988) Origin(s), evolution, and systematics of *Cucurbita pepo* (Cucurbitaceae). *Economic Botany* 42, 4–15.

Decker-Walters, D.S., Walters, T.W., Cowan, C.W. and Smith, B.D. (1993) Isozymic characterization of wild populations of *Cucurbita pepo*. *Journal of Ethnobiology* 13, 55–72.

Delesalle, V.A. and Mooreside, P.D. (1995) Estimating the costs of allocation to male and female functions in a monoecious cucurbit, *Lagenaria siceraria*. *Oecologia* 102, 9–16.

Desai, U.T. and Musmade, A.M. (1998) Pumpkin, squashes and gourds. In: Salunkhe, D.K. and Kadam, S.S. (eds) *Handbook of Vegetable Science and Technology: Production, Composition, Storage and Processing*. Marcel Dekker, New York, pp. 273–298.

Desclaux, D., Ceccarelli, S., Navazio, J., Coley, M., Trouche, G., Aguirre, S., Weltzien, E. and Lancon, J. (2012) Centralized or decentralized breeding: the potentials of participatory approaches for low-input and organic agriculture. In: Lammerts van Bueren, E. and Myers, J.R. (eds) *Organic Crop Breeding*. Wiley-Blackwell, Ames, Iowa, pp. 99–120.

Dey, S.S., Behera, T.K., Munshi, A.D. and Pal, A. (2010) Gynoecious inbred with better combining ability improves yield and earliness in bitter gourd (*Momordica charantia* L.). *Euphytica* 173, 37–47.

Dhillon, N.P.S. and Senapa, N. (2012) Gac means business to a Thai grower-cum-processor. *Feedback from the Field* 15, 3–4.

Dhillon, N.P.S., Sanguansil, S., Schafleitner, R., Wang, Y.-W. and McCreight, J.D. (2016a) Diversity among a wide Asian collection of bitter gourd landraces and their genetic relationships with commercial hybrid cultivars. *Journal of American Society of Horticultural Science* 141, 475–484.

Dhillon, N.P.S., Sanguansil, S., Srimat, S., Cheng, H.C., Lin, C.-C., Srinivasan, R., Kenyon, L., Schafleitner, R., Yang, R.-Y. and Hanson, P. (2016b) Status of cucurbit breeding at AVRDGE-The World Vegetable Center. In: Kozik, E.U. and Paris, H.S. (eds) *Proceedings of Cucurbitaceae 2016, the XIth EUCARPIA Meeting on Genetics and Breeding of Cucurbitaceae*, July 24–28, Warsaw, Poland, pp. 21–25.

Dittmer, H.J. and Talley, B.P. (1964) Gross morphology of tap roots of desert cucurbits. *Botanical Gazette* 125, 121–126.

Dogimont, C., Bordat, D., Pages, C., Boissot, N. and Pitrat, M. (1999) One dominant gene conferring the resistance to the leafminer, *Liriomyza trifolii* (Burgess) Diptera: Agromyzidae in melon (*Cucumis melo* L.). *Euphytica* 105, 63–67.

Duke, J.A. and Ayensu, E.S. (1985) *Medicinal Plants of China, Volume 1*. Reference Publications, Algonac, Michigan.

Durante, M., Lenucci, M.S. and Mita, G. (2014) Supercritical carbon dioxide extraction of carotenoids from pumpkin (*Cucurbita* spp.): a review. *International Journal of Molecular Sciences* 15, 6725–6740.

Endl, J., Achigan-Dako, E.G., Pandey, A.K., Monforte, A.J., Pico, B. and Schaefer, H. (2018) Repeated domestication of melon (*Cucumis melo*) in Africa and Asia and a new close relative from India. *American Journal of Botany* 105, 1662–1671.

Erickson, D.L., Smith, B.D., Clarke, A.C., Sandweiss, D.H. and Tuross, N. (2005) An Asian origin for a 10,000-year-old domesticated plant in the Americas. *Proceedings of the National Academy of Sciences* 51, 18315–18320.

Etchells, J.L., Bell, T.A., Costilow, R.N., Hood, C.E. and Anderson, T.E. (1973) Influence of temperature and humidity on microbial, enzymatic, and physical changes of stored, pickling cucumbers. *Applied Microbiology* 26, 943–950.

Facciola, S. (1990) *Cornucopia: A Source Book of Edible Plants*. Kampong Publications, Vista, California.

FAO (1995) *Production Yearbook for 1994*. No. 48. Food and Agriculture Organization of the United Nations, Rome.

FAOSTAT (2017) FAOSTAT Data. Available at: http://www.fao.org/faostat/en/#data (accessed 9 March 2020).

Fellers, P.J. and Pflug, I.J. (1967) Storage of pickling cucumbers. *Food Technology* 21, 74–78.

Feng, H., Li, X.M., Liu, Z.Y., Wei, P. and Ji, R.Q. (2009) A co-dominant molecular marker linked to the monoecious gene cmACS-7 derived from gene sequence in *Cucumis melo* L. *African Journal of Biotechnology* 8, 3168–3174.

Ferriol, M. and Picó, B. (2008) Pumpkin and winter squash. In: Prohens, J. and Nuez, F. (eds) *Handbook of Plant Breeding. Vegetables, I*. Springer, New York, pp. 317–349.

Fisher, J.B. and Ewers, F.W. (1991) Structural responses to stem injury in vines. In: Putz, F.E. and Mooney, H.A. (eds) *The Biology of Vines*. Cambridge University Press, Cambridge, UK, pp. 99–124.

Floris, E. and Alvarez, J.M. (1995) New sources for powdery mildew resistance in melon from Spanish local cultivars. *Cucurbit Genetics Cooperative Report* 18, 40–42.

Formiga, A.K. and Myers, J.R. (2019) Images and descriptions of *Cucurbita maxima* in Western Europe in the sixteenth and seventeenth centuries. In: Goldman, I. (ed.) *Plant Breeding Reviews, Vol. 43*. John Wiley & Sons, New York, Chapter 9.

Formisano, G., Paris, H.S., Frusciante, L. and Ercolano, M.R. (2010) Commercial Cucurbita pepo squash hybrids carrying disease resistance introgressed from Cucurbita moschata have high genetic similarity. *Plant Genetic Resources* 8(3), 198–203.

Francis, F.J. and Thomson, C.L. (1965) Optimum storage conditions for Butternut squash. *Proceedings of the American Society for Horticultural Science* 86, 451–456.

Frantz, J.D. and Jahn, M.M. (2004) Five independent loci each control monogenic resistance to gummy stem blight in melon (*Cucumis melo* L.). *Theoretical and Applied Genetics* 108, 1033–1038.

Fukino, N., Ohara, T., Sugiyama, M., Kubo, N., Hirao, M., Sakata, Y. and Matsumoto, S. (2012) Mapping of a gene that confers short lateral branching (slb) in melon (*Cucumis melo* L.). *Euphytica* 187, 133–143.

Ganal, M. and Hemleben, V. (1986) Comparison of the ribosomal RNA genes in four closely related Cucurbitaceae. *Plant Systematics and Evolution* 154, 63–77.

Gao, M.L., Zhu, Z.C., Gao, P., and Luan, F.S. (2011) A microsatellite-based genetic map of melon and localization of gene for gynoecious sex expression using recombinant inbred lines. *Acta Horticulturae Sinica* 38(7), 1308–1316.

Garcia-Mas, J., Monoforte, A.J. and Arus, P. (2004) Phylogenetic relationships among *Cucumis* species based on the ribosomal internal transcribed spacer sequence and micro satellite markers. *Plant Systematics and Evolution* 248, 191–203.

Gathman, A.C. and Bemis, W.P. (1990) Domestication of buffalo gourd, *Cucurbita foetidissima*. In: Bates, D.M., Robinson, R.W. and Jeffrey, C. (eds) *Biology and Utilization of the Cucurbitaceae*. Cornell University Press, Ithaca, New York, pp. 335–348.

Gavilánez-Slone, J.M. (2000) Pollination and pollinators of pumpkin and squash (*Cucurbita maxima* Duchesne) grown for seed production in the Willamette Valley of western Oregon. MS thesis, Oregon State University, Corvallis, Oregon. Available at: http://ir.library.oregonstate.edu/concern/graduate_thesis_or_dissertations/db78tf34k

Ge, Y., Li, X., Yang, X.X., Cui, C.S. and Qu, S.P. (2015) Genetic linkage map of *Cucurbita maxima* with molecular and morphological markers. *Genetics and Molecular Research* 14, 5480–5484.

Goffinet, M.C. (1990) Comparative ontogeny of male and female flowers of *Cucumis sativus*. In: Bates, D.M., Robinson, R.W. and Jeffrey, C. (eds) *Biology and Utilization of the Cucurbitaceae*. Cornell University Press, Ithaca, New York, pp. 288–304.

Gong, L., Pachner, M., Kalai, K. and Lelley, T. (2008) SSR-based genetic linkage map of *Cucurbita moschata* and its synteny with *Cucurbita pepo*. *Genome* 51(11), 878–887.

Gonsalves, D., Chee, P., Provvidenti, R., Seem, R. and Slightom, J.L. (1992) Comparison of coat protein-mediated and genetically-derived resistance in cucumbers to infection by cucumber mosaic virus under field conditions with natural challenge inoculations by vectors. *Biotechnology* 10, 1562–1570

Gonsalves, C., Xue, B., Yepes, M., Fuchs, M., Ling, K., Namba, S., Chee, P., Slightom, J.L. and Gonsalves, D. (1994) Transferring cucumber mosaic virus–white leaf strain coat protein gene into *Cucumis melo* L. and evaluating transgenic plants for protection against infections. *Journal of the American Society for Horticultural Science* 119, 345–355.

Gounaris, I., Hardison, R.C. and Boyer, C.D. (1990) Restriction site and genetic map of *Cucurbita pepo* chloroplast DNA. *Current Genetics* 18, 273–275.

Graham, J.D. and Bemis, W.P. (1990) Interspecific trisomics of *Cucurbita moschata*. In: Bates, D.M., Robinson, R.W. and Jeffrey, C. (eds) *Biology and Utilization of the Cucurbitaceae*. Cornell University Press, Ithaca, New York, pp. 421–426.

Grimstad, S.O. and Frimanslund, E. (1993) Effect of different day and night temperature regimes on glasshouse cucumber young plant production, flower bud formation and early yield. *Scientia Horticulturae* 53, 191–204.

Guo, S., Zhang, J., Sun, H., Salse, J., Lucas, W.J. *et al.* (2013) The draft genome of watermelon (Citrullus lanatus) and resequencing of 20 diverse accessions. *Nature Genetics* 45(1), 51–58.

Gürcan, K., Say, A., Yetisir, H. and Denli, N.A. (2015) A study of genetic diversity in bottle gourd [*Lagenaria siceraria* (Molina) Standl.] population, and implication for the historical origins on bottle gourds in Turkey. *Genetic Resources and Crop Evolution* 62, 321–333.

Hall, W.C. (1949) The effects of photoperiod and nitrogen supply on growth and reproduction in the gherkin. *Plant Physiology* 24, 753–769.

Hand, D.W. (1984) Crop responses to winter and summer CO_2 enrichment. *Acta Horticulturae* 162, 45–62.

Hanna, H.Y., Colyer, P.D., Kirkpatrick, T.L., Romaine, D.J. and Vernon, P.R. (1993) Improving yield of cucumbers in nematode-infested soil by double-cropping with a resistant tomato cultivar, using transplants and nematicides. *Proceedings of the Florida State Horticultural Society* 106, 163–165.

Harinantenaina, L., Tanaka, M., Takaoka, S., Oda, M., Mogami, O., Uchida, M. and Asakawa, Y. (2006) *Momordica charantia* constituents and antidiabetic screening of the isolated major compounds. *Chemical and Pharmaceutical Bulletin* 54, 1017–1021.

Hart, J.P. (2004) Can *Cucurbita pepo* gourd seeds be made edible? *Journal of Archaeological Science* 31, 1631–1633.

Hayata, Y., Niimi, Y. and Iwasaki, N. (1995) Synthetic cytokinin – 1-(2-chloro-4-pyridyl)-3-phenylurea (CPPU) – promotes fruit set and induces parthenocarpy in watermelon. *Journal of the American Society for Horticultural Science* 120, 997–1000.

Heiser, C.B. Jr (1979) *The Gourd Book*. University of Oklahoma Press, Norman, Oklahoma.

Herrington, M.E., Prytz, S., Wright, R.M., Walker, I.O., Brown, P., Persley D.M. and Greber, R.S. (2001) 'Dulong QHI' and 'Redlands Trailblazer', PRSV-W-, ZYMV-, and WMV-resistant winter squash cultivars. *HortScience* 36, 811–812.

Hochmuth, R.C. and Hochmuth, G.J. (1991) Nitrogen requirement for mulched slicing cucumbers. *Proceedings of the Soil and Crop Science Society of Florida* 50, 130–133.

Holdsworth, W.L., LaPlant, K.E., Bell, D.C., Jahn, M.M. and Mazourek, M. (2016) Cultivar-based introgression mapping reveals wild species-derived *Pm-0*, the major powdery mildew resistance locus in squash. *PLOS One* 11(12), e0167715.

Holroyd, R. (1914) Morphology and physiology of the axis in Cucurbitaceae. *Botanical Gazette* 78, 1–14.

Hopen, H.J. and Ries, S.K. (1962) The mutually compensating effect of carbon dioxide concentrations and light intensities on the growth of *Cucumis sativus* L. *Proceedings of the American Society for Horticultural Science* 81, 358–364.

Hopp, R.J., Merrow, S.B. and Elbert, E.M. (1960) Varietal differences and storage changes in β-carotene content of six varieties of winter squashes. *Proceedings of the American Society for Horticultural Science* 76, 568–576.

Huang, B., NeSmith, D.S., Bridges, D.C. and Johnson, J.W. (1995) Responses of squash to salinity, waterlogging, and subsequent drainage: II. Root and shoot growth. *Journal of Plant Nutrition* 18, 141–152.

Huang, S., Li, R., Zhang, Z., Li, L., Gu, X., Fan, W., Lucas, WJ., Wang, X., Xie, B., Ni, P. *et al.* (2009) The genome of the cucumber, *Cucumis sativus* L. *Nature Genetics* 41, 1275–1281.

Hughes, D.L. and Yamaguchi, M. (1983) Identification and distribution of some carbohydrates of the muskmelon plant. *HortScience* 18, 739–740.

Hutton, M.G. and Loy, J.B. (1992) Inheritance of Cold Germinability in Muskmelon. *HortScience* 27(7), 826–829.

Hwang, J., Oh, J., Kim, Z., Staub, J.E., Chung, S.M. and Park, Y. (2014) Fine genetic mapping of a locus controlling short internode length in melon (*Cucumis melo* L.). *Molecular Breeding* 34, 949–961.

Islam, S., Munshi, A.D., Mandal, B., Kumar, R. and Behera, T.K. (2010) Genetics of resistance in *Luffa cylindrica* Roem. against Tomato leaf curl New Delhi Virus. *Euphytica* 174, 83–89.

Islam, S., Munshi, A.D., Verma, M., Arya, L., Mandal, B., Behera, T.K., Kumar, R. and Lal, S.K. (2011) Screening of *Luffa cylindrica* Roem. for resistance against Tomato leaf curl New Delhi virus, inheritance of resistance, and identification of SRAP markers linked to the single dominant resistance gene. *Journal of Horticultural Science and Biotechnology* 86, 661–667.

Iwamoto, E., Hayashida, S., Ishida, T. and Morita, T. (2009) Breeding and seasonal adaptability of high-female F_1 hybrid bitter melon (*Momordica charantia* L.) 'Kumaken BP1' using gynoecious inbred line for the seed parent. *Horticultural Research* 8, 143–147.

Jahn, M., Munger, H.M. and McCreight J.D. (2002) Breeding cucurbit crops for powdery mildew resistance. In: Bélanger, R.R., Bushnell, W.R., Dik, A.J. and Carver, T.L. (eds) *The Powdery Mildews: A Comprehensive Treatise.* The American Phytopathological Society, St Paul, Minnesota, pp. 239–248.

Janick, J. and Paris, H.S. (2006) The cucurbit images (1515–1518) of the Villa Farnesina, Rome. *Annals of Botany* 97(2), 165–176.

Jeffrey, C. (1967) Cucurbitaceae. In: Milne-Redhead, E. and Polhill, R.M. (eds) *Flora of Tropical East Africa, Volume 4.* Whitefriars Press, London and Tonbridge, UK.

Jeffrey, C. (1990) Systematics of the Cucurbitaceae: an overview. In: Bates, D.M., Robinson, R.W. and Jeffrey, C. (eds) *Biology and Utilization of the Cucurbitaceae.* Cornell University Press, Ithaca, New York, pp. 3–9.

Jennings, P. and Saltveit, M.E. (1994) Temperature effects on imbibition and germination of cucumber (*Cucumis sativus*) seeds. *Journal of the American Society for Horticultural Science* 119, 464–467.

Jiang, B., Xie, D., Liu, W., Peng, Q. and He, X. (2013) De novo assembly and characterization of the transcriptome, and development of SSR markers in wax gourd (*Benicasa hispida*). *PLOS One* 8(8), e70154.

John, K.J., Khedasana, R., Nissar, V.A.M., Sheen Scariah, Shrikant Sutar, Rao, S.R., Nicar, M.A., Latha, M., Yadav, S.R. and Bhat K.V. (2014) On the occurrence, distribution and taxonomy of *Cucumis setosus* Cogn., an endemic wild edible vegetable from India. *Genetic Resources and Crop Evolution* 61, 345–355.

Karaağaç, O. and Balkaya, A. (2013) Interspecific hybridization and hybrid seed yield of winter squash (*Cucurbita maxima* Duch.) and pumpkin (*Cucurbita moschata* Duch.) lines for rootstock breeding. *Scientia Horticulturae* 149, 9–12.

Karipçin, Z., Sari, N. and Kirnak, H. (2008) Preliminary research on drought resistance of wild and domestic Turkish watermelon accessions. In: Pitrat, M. (ed) Cucurbitaceae 2008, Proceedings of the IXth EUCARPIA Meeting on Genetics and Breeding of Cucurbitaceae, Avignon (France), pp. 493–499.

Kawasaki, S., Miyake, C., Kohchi, T., Fujii, S., Uchida, M. and Yokota, A. (2000) Responses of Wild Watermelon to Drought Stress: Accumulation of an ArgE Homologue and Citrulline in Leaves during Water Deficits. *Plant and Cell Physiology* 41(7), 864–873.

Kays, S.J. and Silva Dias, J.C. (1995) Common names of commercially cultivated vegetables of the world in 15 languages. *Economic Botany* 49, 115–152.

Keinath, A.P., Wintermantel, W.M. and Zitter T.A. (eds) (2017) *Compendium of Cucurbit Diseases and Pests,* 2nd ed. APS Press, St. Paul, Minnesota.

Kihara, H. (1951) Triploid water-melons. *Proceedings of the American Society of Horticultural Science* 58, 217–230.

Kirkbride, J.H. Jr (1993) *Biosystematic Monograph of the Genus Cucumis (Cucurbitaceae).* Parkway Publishers, Boone, North Carolina.

Kistler, L., Montenegro, A., Smith, B.D., Gifford, J.A., Greene, R.E., Newsome, L.A. and Shapiro, B. (2014) Transoceanic drift and the domestication of African bottle gourds in the Americas. *Proceedings of the National Academy of Sciences* 111, 2937–2941.

Kong, W., Chen, N., Liu, T., Zhu, J., Wang, J., He, X. and Jin, Y. (2015) Large-scale transcriptome analysis of cucumber and *Botrytis cinerea* during infection. *PLOS One* 10, e0142221.

Knavel, D.E. (1990) Inheritance of the main dwarf short-internode mutant muskmelon. *HortScience* 25, 1274–1275.

Knavel, D.E. (1991) Productivity and growth of short-internode muskmelon plants at various spacings or densities. *Journal of the American Society for Horticultural Science* 116, 926–929.

Kousik, C.S., Levi, A., Ling, K.-S. and Wechter, W.P. (2008) Potential sources of resistance to cucurbit powdery mildew in US plant introductions of bottle gourd. *HortScience* 43, 1359–1364.

Kousik, C.S., Donahoo, R.S. and Hassell, R. (2012) Resistance in watermelon rootstocks to crown rot caused by *Phytophthora capsici*. *Crop Protection* 39, 18–25.

Kramer, P.J. (1942) Species differences with respect to water absorption at low soil temperatures. *American Journal of Botany* 29, 828–832.

Krug, H. and Liebig, H.P. (1980) Diurnal thermoperiodism of the cucumber. *Acta Horticulturae* 118, 83–94.

Lee, J. (1994) Cultivation of grafted vegetables. I. Current status, grafting methods, and benefits. *HortScience* 29, 235–239.

Levi, A., Thies, J., Ling, K.S., Simmons, Q.M., Kousik, C. and Hassell, R. (2009) Genetic diversity among *Lagenaria siceraria* accessions containing resistance to root-knot nematodes, whiteflies, ZYMV or powdery mildew. *Plant Genetic Resources: Characterization and Utilization* 7, 216–226.

Li, X., An, M., Xiao, Z., Bai, X. and Wu, Y. (2017) Transcriptome analysis of watermelon (*Citrulllus lanatus*) fruits in response to cucumber green mottle mosaic virus (CGMMV) infection. *Scientific Reports* 7, 16747.

Liebig, H.P. (1980) A growth model to predict yield and economical figures of the cucumber crop. *Acta Horticulturae* 118, 165–174.

Ling, J., Mao, Z., Zhai, M., Zeng, F., Yang, Y. and Xie, B. (2017) Transcriptome profiling of *Cucumis metuliferus* infected by *Meloidogyne incognito* provides new insights into putative defense regulatory network in Cucurbitaceae. *Scientific Reports* 7, 3544.

Ling, K.-S., Levi, A., Adkins, S., Kousik, C.S., Hassell, R. and Keinath, A.P. (2013) Development and field evaluation of multiple virus-resistant bottle gourd (*Lagenaria siceraria*). *Plant Disease* 97, 1057–1062.

Lira-Saade, R. (1995) *Estudios Taxonómicos y Ecogeográficos de las Cucurbitaceae Latinoamericanas de Importancia Económica*. Systematic and Ecogeographic Studies on Crop Genepools. 9. International Plant Genetic Resources Institute, Rome, Italy.

Liyanage, R., Nadeeshani, H., Jayathilake, C., Visvanathan, R. and Wimalasiri, S. (2016) Comparative analysis of nutritional and bioactive properties of aerial parts of snake gourd (*Trichosanthes cucumerina* Linn). *International Journal of Food Science* 2016, 8501637. DOI: 10.1155/2016/8501637.

Lopez-Sese, A.I. and Gomez-Guillamon, M.L. (2000) Resistance to cucurbit yellowing stunting disorder virus (CYSDV) in *Cucumis melo* L. *HortScience* 35, 110–113.

Lorenz, O.A. and Maynard, D.N. (1980) *Knott's Handbook for Vegetable Growers*. John Wiley and Sons, New York.

Lower, R.L. and Nienhuis, J. (1990) Prospects for increasing yields of cucumbers via *Cucumis sativus* var. *hardwickii* germplasm. In: Bates, D.M., Robinson, R.W. and Jeffrey, C. (eds) *Biology and Utilization of the Cucurbitaceae*. Cornell University Press, Ithaca, New York, pp. 397–405.

Loy, J.B. and Broderick, C.E. (1990) Growth, assimilate partitioning, and productivity of bush and vine cultivars of *Cucurbita maxima*. In: Bates, D.M., Robinson, R.W. and Jeffrey, C. (eds) *Biology and Utilization of the Cucurbitaceae*. Cornell University Press, Ithaca, New York, pp. 436–447.

Lust, T.A. and Paris, H.S. (2016) Italian horticultural and culinary records of summer squash (*Cucurbita pepo*, Cucurbitaceae) and emergence of the zucchini in 19th-century Milan. *Annals of Botany* 118, 53–69.

Macdougal, D.T. and Spalding, E.S. (1910) *The Water-Balance of Succulent Plants.* Carnegie Institution of Washington, Washington, DC.

MacGillivray, J.H., Hanna, G.C. and Minges, P.A. (1942) Vitamin, protein, calcium, iron, and calorie yield of vegetables per acre and per acre man-hour. *Proceedings of the American Society for Horticultural Science* 41, 293–297.

Macheix, J.-J., Fleuriet A. and Billot J. (2000) *Fruit Phenolics.* CRC Press, Boca Raton, Florida.

McCollum, T.G., Cantliffe, D.J. and Parks, H.S. (1987) Flowering, fruit set, and fruit development in birdsnest-type muskmelons. *Journal of the American Society for Horticultural Science* 112, 161–164.

McCreight, J.D., Nerson, H. and Grumet, R. (1993) Melon *Cucumis melo* L. In: Kalloo, G. and Bergh, B.O. (eds) *Genetic Improvement of Vegetable Crops.* Pergamon Press, New York.

McCreight, J.D., Staub, J.E., Wehner, T.C. and Dhillon, N.P.S. (2013) Gone global: familiar and exotic cucurbits have Asian origins. *HortScience* 48, 1078–1089.

Mendlinger, S. (1994) Effect of increasing plant density and salinity on yield and fruit quality in muskmelon. *Scientia Horticulturae* 57, 41–49.

Merrick, L.C. (1990) Systematics and evolution of a domesticated squash, *Cucurbita argyrosperma*, and its wild and weedy relatives. In: Bates, D.M., Robinson, R.W. and Jeffrey, C. (eds) *Biology and Utilization of the Cucurbitaceae.* Cornell University Press, Ithaca, New York, pp. 77–95.

Merrick, L.C. and Bates, D.M. (1989) Classification and nomenclature of *Cucurbita argyrosperma. Baileya* 23, 94–102.

Metcalf, R.L. and Rhodes, A.M. (1990) Coevolution of the Cucurbitaceae and Luperini (Coleoptera: Chrysomelidae): basic and applied aspects. In: Bates, D.M., Robinson, R.W. and Jeffrey, C. (eds) *Biology and Utilization of the Cucurbitaceae.* Cornell University Press, Ithaca, New York, pp. 167–182.

Miller, C.H. and Wehner, T.C. (1989) Cucumbers. In: Eskin, N.A.M. (ed.) *Quality and Preservation of Vegetables.* CRC Press, Boca Raton, Florida, pp. 245–264.

Minocha, S. (2015) An overview on *Lagenaria siceraria* (Bottle gourd). *Journal of Biomedical and Pharmaceutical Research* 4, 4–10.

Montero-Pau, J., Blanca, J., Esteras, C., Martinez-Perez, E.M., Gomez, P., Monforte, A.J., Canizares, J. and Pico, B. (2017) An SNP-based saturated genetic map and QTL analysis of fruit-related traits in zucchini using genotyping-by-sequencing. *BMC Genomics* 18, 94.

Morgan, W. and Midmore, D. (2002) Bitter melon in Australia. In: *Publication No. 02/134,* p. 29. Rural Industries Research and Development Corporation, Rockhampton, Australia.

Munger, H.M. (1992) The significance of some traits and their combinations in the usage of US cucumber varieties. *Cucurbit Genetics Cooperative Report* 15, 17–18.

Munger, H.M. and Robinson, R.W. (1991) Nomenclature of *Cucumis melo* L. *Cucurbit Genetics Cooperative Report* 14, 43–44.

Munger, H.M., Kyle, M.M. and Robinson, R.W. (1993) Cucurbits. In: *Traditional Crop Breeding Practices: an Historical Review to Serve as a Baseline for Assessing the Role of Modern Biotechnology.* Organisation for Economic Co-operation and Development, Paris, pp. 47–60.

Munshi, A.D., Islam, S., Kumar, R., Mandal, B., Behera, T.K. and Sureja, A.K. (2015) Sponge gourd DSG 6 to combat ToLCNDV. *Indian Horticulture* May–June 2015.

Navazio, J.P. (1994) Utilization of high-carotene cucumber germplasm for genetic improvement of nutritional quality. PhD thesis, University of Wisconsin, Madison, Wisconsin.

Navot, N. and Zamir, D. (1986) Linkage relationships of 19 protein coding genes in watermelon. *Theoretical and Applied Genetics* 72, 274–278.

Navot, N. and Zamir, D. (1987) Isozyme and seed protein phylogeny of the genus *Citrullus* (Cucurbitaceae). *Plant Systematics and Evolution* 156, 61–67.

N'dri, A.N.A., Zoro, B.I.A., Kouamé, L.P., Dumet, D. and Vroh-Bi, I. (2016) On the dispersal of bottle gourd [*Lagenaria siceraria* (Mol.) Standl.] out of Africa: a contribution from the analysis of nuclear ribosomal DNA haplotypes, divergent paralogs and variants of 5.8S protein sequences. *Plant Molecular Biology Reporter* 24, 454–466.

Nerson, H., Cantliffe, D.J., Paris, H.S. and Karchi, Z. (1982) Low-temperature germination of birdsnest-type muskmelons. *HortScience* 17, 639–640.

Newstrom, L.E. (1990) Origin and evolution of chayote, *Sechium edule*. In: Bates, D.M., Robinson, R.W. and Jeffrey, C. (eds) *Biology and Utilization of the Cucurbitaceae*. Cornell University Press, Ithaca, New York, pp. 141–149.

Newstrom, L.E. (1991) Evidence for the origin of chayote, *Sechium edule* (Cucurbitaceae). *Economic Botany* 45, 410–428.

Nijs, A.P.D. den and Visser, D. (1985) Relationships between African species of the genus *Cucumis* L. estimated by the production, vigour and fertility of F_1 hybrids. *Euphytica* 34, 279–290.

Nitsch, J.P., Kurtz, E.B. Jr, Liverman, J.L. and Went, F.W. (1952) The development of sex expression in cucurbit flowers. *American Journal of Botany* 39, 32–43.

Noguera, F.J., Capel, J., Alvarez, J.I. and Lozano, R. (2005) Development and mapping of a codominant SCAR marker linked to the *andromonoecious* gene of melon. *Theoretical and Applied Genetics* 100, 714–720.

Obiagwu, C.J. and Odiaka, N.I. (1995) Fertilizer schedule for yield of fresh fluted pumpkin (*Telfairia occidentalis*) grown in lower Benue river basin of Nigeria. *Indian Journal of Agricultural Sciences* 65, 98–101.

Obrero, A., González-Verdejo, C.I., Die, J.V., Gómez, P., Del Río-Celestino, M. and Román, B. (2013) Carotenogenic gene expression and carotenoid accumulation in three varieties of *Cucurbita pepo* during fruit development. *Journal of Agricultural and Food Chemistry* 61, 6393–6403.

Oliver, M., Garcia-Mas, J., Cardus, M., Pueyo, N., Lopez-Sese, A., Arroyo, M., Gomez-Paniagua, H., Arus, P. and de Vicente, M.C. (2001) Construction of a reference linkage map for melon. *Genome* 44, 836–845.

Orth, A.B., Teramura, A.H. and Sisler, H.D. (1990) Effects of ultraviolet-B on fungal disease development in *Cucumis sativus*. *American Journal of Botany* 77, 1188–1192.

Ortiz-Alamillo, O., Garza-Ortega, S., Sánchez-Estrada, A. and Troncoso-Rojas, R. (2007) Yield and quality of the interspecific cross *Cucurbita argyrosperma* × *C. moschata*. *Cucurbit Genetics Cooperative* 30, 56–59.

Pachner, M. and Lelley, T. (2004) Different genes for resistance to zucchini yellow mosaic virus (ZYMV) in *Cucurbita moschata*. In: Lebeda, A. and H.S. Paris (eds) *Progress in Cucurbit Genetics and Breeding Research*. Proceedings of Cucurbitaceae 2004, the 8th EUCARPIA Meeting on Cucurbit Genetics and Breeding. Palacky University in Olomouc, Czech Republic, pp. 237–243.

Pachner, M., Paris, H.S. and Lelley, T. (2011) Genes for resistance to zucchini yellow mosaic in tropical pumpkin. *Journal of Heredity* 102, 330–335.

Paris, H.S. (1994) Genetic analysis and breeding of pumpkins and squash for high carotene content. In: Linskens, H.-F. and Jackson, J.F. (eds) *Modern Methods of Plant Analysis, Vol. 16: Vegetables and Vegetable Products.* Springer-Verlag, Berlin, pp. 93–115.

Paris, H.S. (2000) Paintings (1769–1774) by AN Duchesne and the history of *Cucurbita pepo*. *Annals of Botany* 85(6), 815–830.

Paris, H.S. (2001) History of the cultivar-groups of *Cucurbita pepo*. *Horticultural Reviews* 25, 71–170.

Paris, H.S. (2008) Summer squash. In: Prohens, J. and Nuez, F. (eds) *Handbook of Plant Breeding. Vegetables, I.* Springer, New York, pp. 351–379.

Paris, H.S. (2012) Semitic-language records of snake melons (*Cucumis melo*, Cucurbitaceae) in the medieval period and the 'piqqus' of the 'faqqous'. *Genetic Resources and Crop Evolution* 59, 31–38.

Paris, H.S. (2015) Origin and emergence of the sweet dessert watermelon, *Citrullus lanatus*. *Annals of Botany* 116, 133–148.

Paris, H.S. and Cohen, S. (2000) Oligogenic inheritance for resistance to zucchini yellow mosaic virus in *Cucurbita pepo*. *Annals of Applied Biology* 136, 209–214.

Paris, H.S. and Padley, L.D. (2014) Gene List for *Cucurbita* species, 2014. Cucurbit Genetics Cooperative. Available at: http://cuke.hort.ncsu.edu/cgc/cgcgenes/gene 14squash.pdf.

Paris, H.S., Daunay, M.C., Pitrat, M. and Janick, J. (2006) First known image of *Cucurbita* in Europe, 1503–1508. *Annals of Botany* 98(1), 41–47.

Paris, H.S., Daunay, M.C. and Janick, J. (2011) Occidental diffusion of cucumber (*Cucumis sativus*) 500–1300 CE: two routes to Europe. *Annals of Botany* 2019, 117–126.

Peet, M.M. and Willits, D.H. (1987) Glasshouse CO_2 enrichment alternatives: effects of increasing concentration or duration of enrichment on cucumber yields. *Journal of the American Society for Horticultural Science* 112, 236–241.

Perl-Treves, R. and Galun, E. (1985) The *Cucumis* plastome: physical map, intrageneric variation and phylogenetic relationships. *Theoretical and Applied Genetics* 71, 417–429.

Perry, K.B. and Wehner, T.C. (1990) Prediction of cucumber harvest date using a heat unit model. *HortScience* 25, 405–406.

Pfister, S.C., Eckerter, P.W., Schirmel, J., Cresswell, J.E. and Entling, M.H. (2017) Sensitivity of commercial pumpkin yield to potential decline among different groups of pollinating bees. *Royal Society of Open Science* 4(5), 170102. DOI: 10.1098/rsos.170102.

Pier, J.W. and Doerge, T.A. (1995) Concurrent evaluation of agronomic, economic, and environmental aspects of trickle-irrigated watermelon production. *Journal of Environmental Quality* 24, 79–86.

Pierce, L.C. (1987) *Vegetables. Characteristics, Production, and Marketing.* John Wiley and Sons, New York.

Pierce, L.K. and Wehner, T.C. (1990) Review of genes and linkage groups in cucumber. *HortScience* 25, 605–615.

Piperno, D.R. and Stothert, K.E. (2003) Phytolith evidence for early Holocene *Cucurbita* domestication in southwest Ecuador. *Science* 299, 1054–1057.

Pitrat, M. (1994) Gene list for *Cucumis melo* L. *Cucurbit Genetics Cooperative Report* 17, 135–147.

Prakash, K., Pati, K., Arya, L., Pandey, A. and Verma, M. (2014) Population structure and diversity in cultivated and wild *Luffa* species. *Biochemical Systematics and Ecology* 56, 165–170.

Pratt, H.K., Goeschl, J.D. and Martin, F.W. (1977) Fruit growth and development, ripening, and the role of ethylene in the 'Honey Dew' muskmelon. *Journal of the American Society for Horticultural Science* 102, 203–210.

Provvidenti, R. (1990) Viral diseases and genetic sources of resistance in *Cucurbita* species. In: Bates, D.M., Robinson, R.W. and Jeffrey, C. (eds) *Biology and Utilization of the Cucurbitaceae*. Cornell University Press, Ithaca, New York, pp. 427–435.

Provvidenti, R. (1991) Inheritance of Resistance to the Florida Strain of Zucchini Yellow Mosaic Virus in Watermelon. *HortScience* 26(4), 407–408.

Provvidenti, R., Robinson, R.W. and Munger, H.M. (1978a) Resistance in feral species to six viruses infecting *Cucurbita*. *Plant Disease Reporter* 62, 326–329.

Provvidenti, R., Robinson, R.W. and Munger, H.M. (1978b) Multiple virus resistance in *Cucurbita* species. *Cucurbit Genetics Cooperative Report* 1, 26–27.

Puchalski, J.T. and Robinson, R.W. (1990) Electrophoretic analysis of isozymes in *Cucurbita* and *Cucumis* and its application for phylogenetic studies. In: Bates, D.M., Robinson, R.W. and Jeffrey, C. (eds) *Biology and Utilization of the Cucurbitaceae*. Cornell University Press, Ithaca, New York, pp. 60–76.

Quemada, H.D. and Groff, D.W. (1995) Genetic engineering approaches in the breeding of virus resistant squash. In: Lester, G. and Dunlap, J. (eds) *Cucurbitaceae '94*. Gateway Printing, Edinburg, Texas, pp. 93–94.

Rai, P.K., Jaiswal, D., Singh, R.K., Gupta, R.K. and Watal, G. (2009) Glycemic properties of *Trichosanthes silica* leaves. *Pharmaceutical Biology* 46, 894–899.

Ramachandran, C. and Seshadri, V.S. (1986) Cytological analysis of the genome of cucumber (*Cucumis sativus* L.) and muskmelon (*Cucumis melo* L.). *Zeistschrift Pflanzenzuchtung* 96, 25–38.

Resh, H.M. (1987) *Hydroponic Food Production*. Woodbridge Press, Santa Barbara, California.

Reyes, M.E.C., Gildemache, B.H. and Jansen, G.J. (1994) *Momordica* L. In: Piluek (ed.) *Plant Resources of South-East Asia, No. 8, Vegetables*. Pudoc Scientific Publishers, Wageningen, pp. 206–210.

Rhodes, B. and Zhang, X. (1995) Gene list for watermelon. *Cucurbit Genetics Cooperative Report* 18, 69–84.

Richardson, J.B. (1972) The pre-Columbian distribution of the bottle gourd (*Lagenaria sicerarid*): a re-evaluation. *Economic Botany* 26, 265–273.

Robinson, R.W. (1992) Genetic resistance in the Cucurbitaceae to insects and spider mites. *Plant Breeding Reviews* 10, 309–360.

Robinson, R.W. (2010–2011) Pollination of squash before and after the day of anthesis. *Cucurbit Genetics Cooperative Report* 33–34, 51–52.

Robinson, R.W. and Shail, J.W. (1987) Genetic variability for compatibility of an interspecific cross. *Cucurbit Genetics Cooperative Report* 10, 88–89.

Robinson, R.W., Shannon, S. and Guardia, M.D. de la (1969) Regulation of sex expression in the cucumber. *BioScience* 19, 141–142.

Robinson, R.W., Whitaker, T.W. and Bohn, G.W. (1970) Promotion of pistillate flowering in *Cucurbita* by 2-chloroethylphosphonic acid. *Euphytica* 19, 180–183.

Robinson, R.W., Cantliffe, D.J. and Shannon, S. (1971) Morphactin-induced partheno-carpy in the cucumber. *Science* 171, 1251–1252.

Rordan var Eysinga, J.P.N.L. and Smilde, K.W. (1969) *Nutritional Disorders in Cucumbers and Gherkins Under Glass.* Centre for Agricultural Publishing and Documentation, Wageningen, Holland.

Roy, R.P. and Saran, S. (1990) Sex expression in the Cucurbitaceae. In: Bates, D.M., Robinson, R.W. and Jeffrey. C. (eds) *Biology and Utilization of the Cucurbitaceae.* Cornell University Press, Ithaca, New York, pp. 251–268.

Rudich, J. (1990) Biochemical aspects of hormonal regulation of sex expression in cucurbits. In: Bates. D.M., Robinson, R.W. and Jeffrey. C. (eds) *Biology and Utilization of the Cucurbitaceae.* Cornell University Press, Ithaca, New York, pp. 269–280.

Rundel, P.W. and Franklin, T. (1991) Vines in arid and semi-arid ecosystems. In: Putz, F.E. and Mooney. H.A. (eds) *The Biology of Vines.* Cambridge University Press, Cambridge, pp. 337–356.

Saade, R.L. (1996) Chayote, *Sechium edule* (Jacq.) Sw. In: *Promoting the Conservation and Use of Underutilized and Neglected Crops. 8.* Institute of Plant Genetics and Crop Plant Research, Gatersleben/International Plant Genetic Resources Institute, Rome, p.58.

Schaefer, H. and Renner, S.S. (2011) Phylogenetic relationships in the order Cucurbitales and a new classification of the gourd family (Cucurbitaceae). *Taxon* 60, 122–138.

Schaefer, H., Heibl, C. and Renner, S.S. (2009) Gourds afloat: a dated phylogeny re-veals an Asian origin of the gourd family (Cucurbitaceae) and numerous oversea dispersal events. *Proceedings of the Royal Society B: Biological Sciences* 276, 843–851.

Schales, F.D. and Isenberg, F.M. (1963) The effect of curing and storage on chemical composition and taste acceptability of winter squash. *Proceedings of the American Society for Horticultural Science* 83, 667–674.

Sebastian, P.M., Schaefer, H., Telford, I.H.R. and Renner, S.S. (2010) Cucumber and melon have their wild progenitors in India, and the sister species of *Cucumis melo* is from Australia. *Proceedings of the National Academy of Sciences USA* 107, 14269–14273.

Seshadri, V.S. (1986) Cucurbits. In: Bose, T.K. and Som, M.G. (eds) *Vegetable Crops in India.* Naya Prokash, Calcutta, India, pp. 91–164.

Seshadri, V.S. and More, T.A. (2009) *Cucurbit Vegetables – Biology, Production and Utilization.* Studium Press, New Delhi.

Shannon, S. and Robinson, R.W. (1979) The use of ethephon to regulate sex expression of summer squash for hybrid seed production. *Journal of the American Society of Horticultural Science* 104, 674–677.

Shannon, M.C. and Grieve, C.M. (1998) Tolerance of vegetable crops to salinity. *Scientia Horticulturae* 78(1-4), 5–38.

Sharples, G.C. and Foster, R.E. (1958) The growth and composition of cantaloupe plants in relation to the calcium saturation percentage and nitrogen level of the soil. *Proceedings of the American Society for Horticultural Science* 72, 417–425.

Sherman, M., Paris, H.S. and Allen, J.J. (1987) Storability of summer squash as affected by gene B and genetic background. *HortScience* 22, 920–922.

Shih, S. (1962) *A Preliminary Survey of the Book Ch'i Min Yao Shu: An Agricultural Encyclopaedia of the 6th Century.* 2nd edn. Science Press, Peking, China.

Singh, A.K. (1990) Cytogenetics and evolution in the Cucurbitaceae. In: Bates, D.M., Robinson, R.W. and Jeffrey, C. (eds) *Biology and Utilization of the Cucurbitaceae.* Cornell University Press, Ithaca, New York, pp. 10–28.

Singh, A.K., Singh, R., Weeden, N.F., Robinson, R.W. and Singh, N.K. (2011) A linkage map for *Cucurbita maxima* based on Randomly Amplified Polymorphic DNA (RAPD) markers. *Indian Journal of Horticulture* 68(1), 44–50.

Singh, B.K., Ramakrishna, Y. and Verma, V.K. (2015) Chow-chow (*Sechium edule*): best alternative to shifting cultivation in Mizoram. *Indian Journal of Hill Farming* 28, 158–161.

Singh, D. and Dathan, A.S.R. (1990) Seed coat anatomy of the Cucurbitaceae. In: Bates, D.M., Robinson, R.W. and Jeffrey, C. (eds) *Biology and Utilization of the Cucurbitaceae.* Cornell University Press, Ithaca, New York, pp. 225–238.

Singh, N.P. and Matta, N.K. (2010) Levels of seed proteins in *Citrullus* and *Praecitrullus* accessions. *Plant Systematics and Evolution* 290, 47–56.

Singh, S.P. (2013) *Cucurbits – Biodiversity, Breeding and Production in Uttar Pradesh.* Uttar Pradesh State Biodiversity Board, Lucknow, India.

Singogo, W., Lamont, W.J. Jr and Marr, C.W. (1996) Fall-planted cover crops support good yields of muskmelons. *HortScience* 31, 62–64.

Sinnott, E.W. (1932) Shape changes during fruit development in *Cucurbita* and their importance in the study of shape inheritance. *The American Naturalist* 66, 301–309.

Slack, G. and Hand, D.W. (1983) The effect of day and night temperatures on the growth, development and yield of glasshouse cucumbers. *Journal of Horticultural Science* 58, 567–573.

Smith, B.D. (1997a) The initial domestication of *Cucurbita pepo* in the Americas 10,000 years ago. *Science* 276, 932–934.

Smith, B.D. (1997b) Reconsidering the Ocampo caves and the era of incipient cultivation in Mesoamerica. *Latin American Antiquity* 8, 342–383.

Smith, B.D. (2006) Eastern North America as an independent center of plant domestication. *Proceedings of the National Academy of Sciences* 103, 12223–12228.

Soltani, N., Anderson, J.L. and Hamson, A.R. (1995) Growth analysis of watermelon plants grown with mulches and rowcovers. *Journal of the American Society for Horticultural Science* 120, 1001–1009.

Sood, A., Kaur, P. and Gupta, R. (2012) Phytochemical screening and antimicrobial assay of various seeds extract of Cucurbitaceae family. *International Journal of Applied Biology and Pharmaceutical Technology* 3, 401–409.

Sousa, A., Bellot, S., Fuchs, J., Houben, A. and Renner, S.S. (2016) Analysis of transposable elements and organelles DNA in male and female genomes of a species with a huge Y chromosome reveals distinct Y centromeres. *The Plant Journal* 88, 387–396.

Stocking, K.M. (1955) Some taxonomic and ecological considerations of the genus *Marah* (Cucurbitaceae). *Madroño* 13(4), 113–144.

Sun, X., Wang, Z., Gu, Q., Li, H., Han, W. and Shi, Y. (2017) Transcriptome analysis of *Cucumis sativus* infected by cucurbit chlorotic yellows virus. *Virology Journal* 14, 18.

Swiader, J.M., Sipp, S.K. and Brown, R.E. (1994) Pumpkin growth, flowering, and fruiting response to nitrogen and potassium sprinkler fertigation in sandy soil. *Journal of the American Society for Horticultural Science* 119, 414–419.

Tapley, W.T., Enzie, W.D. and van Eseltine, G.P. (1937) *The Vegetables of New York. Volume 1, Part IV: The Cucurbits.* New York State Agricultural Experiment Station, Geneva.

Telford, R.H., Schaefer, H., Greuter, W. and Renner, S.S. (2011) A new Australian species of *Luffa* (Cucurbitaceae) and typification of two Australian *Cucumis* names, all based on specimens collected by Ferdinand Mueller in 1856. *Phyto Keys* 5, 21–29.

Tian, S., Jian, L., Goa, Q., Zhang, J., Zong, M., Zhang, H., Ren, Y., Gun, S., Gong, G. Liu, F. and Xu, Y. (2017) Efficient CRISPR/Cas9-based gene knockout in watermelon. *Plant Cell Reports* 26, 399–406.

Trehane, P., Brickell, C.D., Baum, B.R., Hetterscheid, W.L.A., Leslie, A.C., McNeill, J., Spongberg, S.A. and Vrugtman. F. (eds) (1995) *International Code of Nomenclature for Cultivated Plants – 1995.* Quarterjack Publishing, Wimborne, UK.

US NPGS GRIN Taxonomy (2017) GRIN-Global Species Data. Available at: https://npgsweb.ars-grin.gov/gringlobal/taxon/taxonomysearch.aspx (accessed 9 March 2020).

USDA Nutrient Database (2014) USDA Database for the Flavonoid Content of Selected Foods. Release 3.1 (May 2014). Available at: https://www.ars.usda.gov/ARSUserFiles/80400525/Data/Flav/Flav_R03-1.pdf (accessed 1 July 2017).

Wall, J.R. and York. T.L. (1960) Genetic diversity as an aid to interspecific hybridization in *Phaseolus* and in *Cucurbita. Proceedings of the American Society for Horticultural Science* 75, 419–428.

Walters, S.A. and Wehner, T.C. (1994) Evaluation of the US cucumber germplasm collection for early flowering. *Plant Breeding* 112, 234–238.

Walters, S.A., Wehner, T.C. and Barker, K.R. (1993) Root-knot nematode resistance in cucumber and horned cucumber. *HortScience* 28, 151–154.

Walters, T.W. and Decker-Walters, D.S. (1988) Balsam-pear (*Momordica charantia,* Cucurbitaceae). *Economic Botany* 42, 286–288.

Waminal, N.E. and Kim, H.H. (2012) Dual-color FISH karyotype and rDNA distribution analysis on four Cucurbitaceae species. *Horticulture, Environment, and Biotechnology* 53, 49–56.

Waminal, N.E., Kim, N.S. and Kim, H.H. (2011) Dual-color FISH karyotype analyses using rDNAs in three Cucurbitaceae species. *Genes and Genomics* 33, 521.

Wang, S., Yang, X., Xu, M., Lin, X., Lin, T., Qi, J., Shao, G., Tian, N., Yang, Q. and Zhang, Z. (2015) A rare SNP identified a TCP transcription factor essential for tendril development in cucumber. *Molecular Plant* 8, 1795–1808.

Weeden, F. (1984) Isozyme studies indicated that the genus *Cucurbita* is an ancient tetraploid. *Cucurbit Genetics Cooperative Report* 7, 84–85.

Weeden, N.F. and Robinson, R.W. (1986) Allozyme segregation ratios in the interspecific cross *Cucurbita maxima* × *C. ecuadorensis* suggest that hybrid breakdown is not caused by minor alterations in chromosome structure. *Genetics* 114, 593–609.

Wehner, T.C. (1993) Gene list update for cucumber. *Cucurbit Genetics Cooperative Report* 16, 92–97.

Wehner, T. C. (1999) Heterosis in vegetable crops. In: Coors, J.G. and Pandey, S. (eds) *Genetics and Exploitation of Heterosis in Crops.* American Society of Agronomy, Madison, Wisconsin, pp. 387–397.

Wehner, T.C. and Ellington, T.L. (1995) Growth regulator effects on sex expression of luffa sponge gourd. *Cucurbit Genetics Cooperative Report* 18, 68.

Wehner, T.C. and Humphries, E.G. (1995) A single-fruit seed extractor for cucumbers. *HortTechnology* 5, 268–273.

Wessel-Beaver, L. (2000) *Cucurbita argyrosperma* sets fruits in fields where *C. moschata* is the only pollen source. *Cucurbit Genetics Cooperative Report* 23, 62–63.

Wessel-Beaver, L., Cuevas, H.E., Andres, T.C. and Piperno, D.R. (2004) Genetic compatibility between *Cucurbita moschata* and *C. argyrosperma*. In: Lebeda, A. and Paris, H.S. (eds) *Progress in Cucurbit Genetics and Breeding Research*. Proceedings of Cucurbitaceae 2004, the 8th EUCARPIA Meeting on Cucurbit Genetics and Breeding. Palacky University in Olomouc, Czech Republic, pp. 393–400.

Whitaker, T.W. and Davis, G.N. (1962) *Cucurbits. Botany, Cultivation, and Utilization*. Interscience Publishers, New York.

Whitaker, T.W. and Robinson, R.W. (1986) Squash breeding. In: Bassett, M.J. (ed.) *Vegetable Breeding*. AVI Publishing Company, Westport, Connecticut, pp. 209–242.

White, J.C. (2002) Differential bioavailability of field-weathered p, p′-DDE to plants of the *Cucurbita* and *Cucumis* genera. *Chemosphere* 49, 143–152.

Widders, I.E. and Price, H.C. (1989) Effects of plant density on growth and biomass partitioning in pickling cucumbers. *Journal of the American Society for Horticultural Science* 114, 751–755.

Wien, H.C. (1997) The cucurbits: cucumber, melon, squash and pumpkin. In: Wien, H.C. (ed.) *The Physiology of Vegetable Crops*. CAB International, Wallingford, UK.

Wilson, H.D., Doebley, J. and Duvall, M. (1992) Chloroplast DNA diversity among wild and cultivated members of *Cucurbita* (Cucurbitaceae). *Theoretical and Applied Genetics* 84, 859–865.

Wu, S., Shamimuzzaman, M., Sun, H., Salse, J., Sui, X., Wilder, Al., Wu, Z., Levi, A., Xu, Y., Ling, K. and Fei, Z. (2017) The bottle gourd genome provides insights into Cucurbitaceae evolution and facilitates mapping of a papaya ring-spot virus resistance locus. *The Plant Journal* 92, 963–975.

Wyatt, L.E., Strickler, S.R., Mueller, L.A. and Mazourek, M. (2016) Comparative analysis of *Cucurbita pepo* metabolism throughout fruit development in acorn squash and oilseed pumpkin. *Horticulture Research* 3, 16045.

Xu, J.P. (2017) *Cancer Inhibitors from Chinese Natural Medicines*. CRC Press, Boca Raton, Florida.

Xu, X., Ji, J., Xu, Q., Qi, X., Weng, Y. and Chen, X. (2018) The major-effect quantitative trait locus CsARN6.1 encodes an AAA ATPase domain-containing protein that is associated with waterlogging stress tolerance by promoting adventitious root formation. *The Plant Journal* 93, 917–930.

Yang, R.Z. and Tang, C.S. (1988) Plants used for pest control in China: a literature review. *Economic Botany* 42, 376–406.

Yang, S.L. and Walters, T.W. (1992) Ethnobotany and the economic role of the Cucurbitaceae of China. *Economic Botany* 46, 349–367.

Yasuda, M., Iwamoto, M., Okabe, H. and Yamauchi, T. (1984) Structure of momordicines I, II, and III, the bitter principles in the leaves and vines of *Momordica charantia* L. *Chemical Pharmaceutical Bulletin* 32, 3044–2047.

Yetişir, H. and Uygur, V. (2010) Responses of grafted watermelon onto different gourd species to salinity stress. *Journal of Plant Nutrition* 33, 315–327.

Yongan, C., Bingkui, Z., Enhui, Z. and Zunlian, Z. (2002) Germplasm innovation by interspecific crosses in pumpkin. *Cucurbit Genetics Cooperative Report* 25, 56–57.

York, A. (1992) Pests of cucurbit crops: marrow, pumpkin, squash, melon and cucumber. In: McInlar, R.G.C. (ed.) *Vegetable Crop Pests*. CRC Press, Boca Raton, Florida, pp. 139–161.

Zhang, G., Ren, Y., Sun, H., Guo, S., Zhang, F., Zhang, J., Zhang, H., Jia, Z., Fci. Z., Xu,Y. and Li, H. (2015) A high-density genetic map for anchoring genome sequences and identifying QTLs associated with dwarf vine in pumpkin (*Cucurbita maxima* Duch.). *BMC Genomics* 16(1), 1101.

Zhang, X. and Jiang, Y. (1990) Edible seed watermelons (*Citrullus lanatus* (Thunb.) Matsum. & Nakai) in Northwest China. *Cucurbit Genetics Cooperative Report* 13, 40–42.

Zhang, H., Gong, G., Guo, S., Ren, Y., Xu, Y. and Ling, K.-S. (2011) Screening the USDA Watermelon Germplasm Collection for Drought Tolerance at the Seedling Stage. *HortScience* 46(9), 1245–1248.

Zhao, J.L., Pan, J.S., Guan, Y., Nie, J.T., Yang, J.J., Qu, M.L., He, H.L. and Cai, R. (2015) Transcriptome analysis in *Cucumis sativus* identifies genes involved in multicellular trichome development. *Genomics* 105, 296–303.

Zitter, T.A., Hopkins, D.L. and Thomas, C.E. (eds) (1996) *Compendium of Cucurbit Diseases*. APS (American Phytopathological Society) Press, St Paul, Minnesota.

Zraidi, A., Stift, G., Pachner, M., Shojaeiyan, A., Gong, L. and Lelley, T. (2007) A consensus map for *Cucurbita pepo*. *Molecular Breeding* 20(4), 375–388.

INDEX

Note: Page numbers in **bold** type refer to **figures**
Page numbers in *italic* type refer to *tables*

CABI – who we are and what we do

This book is published by **CABI**, an international not-for-profit organisation that improves people's lives worldwide by providing information and applying scientific expertise to solve problems in agriculture and the environment.

CABI is also a global publisher producing key scientific publications, including world renowned databases, as well as compendia, books, ebooks and full text electronic resources. We publish content in a wide range of subject areas including: agriculture and crop science / animal and veterinary sciences / ecology and conservation / environmental science / horticulture and plant sciences / human health, food science and nutrition / international development / leisure and tourism.

The profits from CABI's publishing activities enable us to work with farming communities around the world, supporting them as they battle with poor soil, invasive species and pests and diseases, to improve their livelihoods and help provide food for an ever growing population.

CABI is an international intergovernmental organisation, and we gratefully acknowledge the core financial support from our member countries (and lead agencies) including:

Discover more

To read more about CABI's work, please visit: **www.cabi.org**

Browse our books at: **www.cabi.org/bookshop**,
or explore our online products at: **www.cabi.org/publishing-products**

Interested in writing for CABI? Find our author guidelines here:
www.cabi.org/publishing-products/information-for-authors/